Techniques for Ship Handling and Bridge Team Management

Techniques for Ship Handling and Bridge Team Management

Hiroaki Kobayashi

Routledge
Taylor & Francis Group

LONDON AND NEW YORK

First edition published 2020 by Routledge/Productivity Press

2 Park Square, Milton Park, Abingdon, Oxon OX14 4RN
605 Third Avenue New York, NY 10017

Routledge is an imprint of the Taylor & Francis Group, an informa business

First issued in paperback 2021

Publisher's Note

The publisher has gone to great lengths to ensure the quality of this reprint but points out that some imperfections in the original copies may be apparent.

ISBN-13: 978-0-367-31325-8 (hbk)
ISBN-13: 978-1-03-217653-6 (pbk)
DOI: 10.1201/9780429316272

Contents

About the Author   xi
A Note from the Author   xiii
Introduction   xv

PART I
Techniques for Ship Handling   I

Preface   3

1 Factors in Achieving Safe Navigation   5

    1.1   *Difficulty of the Navigational Environment 5*
    1.2   *Factors Affecting Navigational*
         Difficulty 7
         Examples of Rules of Navigation Contributing
            to Ship Navigation Safety 9
         Competencies of Seafarers to Overcome
            Environmental Difficulty 10
    1.3   *Ship Handling Competency of Seafarers 12*
    1.4   *Conditions Necessary for Safe Navigation 15*
    1.5   *Techniques Necessary for Safe Navigation 17*
    1.6   *Applicable Navigational Situations 21*

2 Analysis of Techniques for Ship Handling   23

    2.1   *Technique of Passage Planning 31*
         Definition 31
         Functions 31
         Factors affecting the achievement of functions for
            passage planning 32

Making passage plans 33
Example of nautical chart descriptions when navigating
narrow waters 35

2.2 *Technique of Lookout 38*
Definition 38
Functions 38
Factors affecting the achievement of lookout functions 38
Performing lookout 40

2.3 *Technique of Position Fixing 47*
Definition 47
Functions 47
Factors affecting the achievement of position-fixing functions 48

2.4 *Technique of Maneuvering 50*
Definition 50
Functions 50
Factors affecting the achievement of maneuvering functions 53

2.5 *Technique of Observing Rules of Navigation*
and Other Laws and Regulations 55
Definition 55
Functions 55
Factors affecting the achievement of law and
regulatory observance functions 56

2.6 *Technique of Communication 57*
Definition 57
Functions 57
Factors affecting the achievement of
communication technique 58

2.7 *Technique of Instrument Operation 61*
Definition 61
Functions 61
Factors affecting the achievement of techniques
of instrument operation 66
Notes 66

2.8 *Technique of Handling Emergencies 68*
Definition 68
Functions 68
Technique of preparation for emergency situations 70

2.9 *Technique of Management: Managing*
Techniques and Team Activity 71
Definition 71
Technical management 71

Team management 72
Learning the management technique 74
Functions included in the technique of management 75
References 78

3 Inadequate Knowledge and Competency Often Observed
   in Inexperienced Seafarers   79

   3.1   *Characteristics of Inadequate Action in Planning 80*
   3.2   *Characteristics of Inadequate Action in Lookout 81*
   3.3   *Characteristics of Inadequate Action in Position Fixing 83*
   3.4   *Characteristics of Inadequate Action in Maneuvering 85*
   3.5   *Characteristics of Inadequate Action in
          Observing Laws and Regulations 87*
   3.6   *Characteristics of Inadequate Action in Communication 89*
   3.7   *Characteristics of Inadequate Action
          in Instrument Operation 91*
   3.8   *Characteristics of Inadequate Action in Management 94*
   References 96

4 Significance and Use of Elemental Technique Development   97

   4.1   *Significance of Elemental Technique Development 97*
   4.2   *Techniques and Competency 101*
   4.3   *Limits of Seafarer Competency and
          Competency Expansion 105*
   Reference 108

Postscript   109

PART II
Bridge Team Management   111

Preface   113

5 Techniques Necessary for Safe Navigation   115

6 Factors in Achieving Safe Navigation   121

   6.1 *Difficulty of the Navigational Environment 122*
   6.2 *Seafarer Competency in Ship Handling 125*
   6.3 *Conditions Necessary for Safe Navigation 127*
   6.4 *Role of the Bridge Team in Ensuring Safe Navigation 129*

7 Background of Bridge Team Management   131

  7.1 Definition of Bridge Resource Management and
          Bridge Team Management (Kobayashi 2012) 132
  7.2 Introduction to Team Management in Aircraft Handling 134
  7.3 Introduction to Team Management in Ship Handling 138
          Accident analysis 143
          Chronological sequence of problems 143
  References 145

8 Bridge Team Management   147

  8.1 Necessity of Bridge Team Management Training 148
  8.2 Reasons for Organizing a Bridge Team 151
  8.3 Special Aspects and Necessary Functions of Teamwork 152
  8.4 Communication 154
  8.5 Cooperation 159
  8.6 Necessary Functions of a Team Leader 162
  8.7 Examples of Team Activity Implementation 165
  8.8 Captain's Briefing 166
  8.9 Methods of Communication 168
          Communication for lookout duties 169
          Communication for position-fixing duties 172
  8.10 Necessary Conditions for Motivating
          Teamwork Observed in Actual Cases 174
          Necessary Conditions to Motivate Teams 178
  8.11 Effective Use of Resources 179
  8.12 Summary: Necessary Competencies
          for Bridge Team Management 181

Postscript   183

PART III
Bridge Team Management/Bridge Resource
Management Training   185

Preface   187

9 Training System   189

  9.1 Summary 190
  9.2 Objectives of Education and Training 191

9.3 *Conditions Necessary for Achieving the Objectives*
    of Bridge Team Management Training 192
9.4 *Training System 194*

10 Bridge Team Management Training Structure   199

10.1 *Course Timetable 199*
10.2 *Details of Education and Training 203*

11 Bridge Team Management Training Examples   211

11.1 *Implementing Training 211*
    Assessment items corresponding to possible events 213
11.2 *Examples of Exercises Using Ship Handling Simulator 215*

Postscript   223
Summary of Key Factors   225
Index   241

About the Author

Professor Emeritus Dr. Hiroaki Kobayashi

- Professor emeritus of Tokyo University of Marine Science and Technology
- Doctoral degree from University of Tokyo
- Master's degree from Hiroshima University
- Graduate of Tokyo University of Mercantile Marine
- Graduate of Osaka University

Academic activities on maritime technology

- President of Japan Institute of Navigation (2004–2006)
- Chairman and area representative of Asian and Pacific region at the International Marine Simulator Forum (2002–2009)
- Representative of the Asian Conference on Marine System and Safety Research (2001–2014)
- Member of Steering Committee of the International Navigation Simulator Lecturers' Conference at the International Maritime Lecturers Association (1980–2016)

Activities on maritime education and training

- Development of the following model courses, certified by the International Classification Society as Class NK:
  (1) Model course for Maritime Education and Training on Bridge Team Management and Bridge Resource Management
  (2) Model course for Maritime Education and Training for Instructor Using Ship Handling Simulator
- Presided over the following training courses for mariners of shipping companies:
  (1) Training course on bridge team management and bridge resource management

(2) Training course on ship handling for berthing jetty under tug-boats support
(3) Training course for freshmen
(4) Training course for young officers

- Conducted instructor training courses for captains from shipping companies and instructors from the following countries: Japan, Vietnam, Indonesia, China, Turkey, the United Kingdom, Malaysia, the Philippines, and India.

A Note from the Author

This book consists of three parts. Prior to "Part I, Techniques for Ship Handling", the "Introduction" outlines important issues as a preliminary guide for readers of this book. The "Introduction" allows readers to gain a background view of "clarifying the role of humans in system operation", which this book is aiming at. This view is a new one, not found in previous publications.

This book aims to provide an overall explanation of the techniques demanded of seafarers while engaged in handling a ship. Accordingly, the necessary techniques for achieving safe navigation, which have been developed by seafarers over a long period of time, have been organized to show the importance of those techniques that seafarers need to accomplish in actual situations. It is hoped that the systematization introduced in this book is taking a step forward in clarifying the overall structure of the techniques for achieving safe navigation.

In Part I, the book initially considers the necessary conditions for maintaining safety, which is the primary objective of ship navigation. This is based on the relationship between the techniques employed by seafarers and the environments requiring seafarers to implement these techniques. Next, the techniques that seafarers are expected to use are analyzed. The results of the analysis reveal that all the techniques necessary for safe navigation are organized into aggregates of nine elemental techniques. This part explains the details of each of these nine elemental techniques and shows how such elemental techniques need to be executed in actual navigation. Furthermore, Chapter 4, in the last section of Part I, presents the aforementioned new approach to systems being resolved into necessary elemental techniques, which could provide an effective solution for existing problems in related fields.

In Part II of the book, the concept of team management, which has been specially noted in recent years, is dealt with. Especially, the significance of team activities for safe navigation and the conditions that are necessary for a team to perform the necessary functions are explained. These team management functions are necessary not only in ship navigation but also for teams to achieve specific objectives in other fields. Thus, it is hoped that for those in other fields requiring teamwork, this book will provide valuable insights.

In Part III of the book, details of training to develop team management competency carried out in the last 15 years are introduced. The training introduced here is employed as the accreditation criteria for training in Bridge Team Management by the Nippon Kaiji Kyokai (Class NK): International Classification Society. Those requiring detailed information on the training introduced here are recommended to contact the Ship Maneuvering Simulator Committee of the Japan Institute of Navigation.

The viewpoint of maritime techniques taken in this book differs in many ways from that of past publications. Therefore, the key factors have been summarized at the end of each section of the main text in order to ensure a greater understanding of its contents. As a matter of course, these factors correspond to the necessary behavior in actual situations onboard. Accordingly, for onboard use, those factors are gathered together in a catalogue shown at the end of this book. It is hoped that this book will be helpful in providing beneficial insight into safe navigation to all readers, from novice to experienced.

Hiroaki Kobayashi

Introduction

This book explains the techniques necessary for safe navigation. Although several works have been published on these techniques, they lack detailed explanations, and all ship handling techniques have been expressed in terms of seamanship, without systematic discussion of the functions involved in ship handling or the detailed discussion of each of these functions.

As a student, the author studied the development of ship maneuvering simulators, which were unavailable in Japan at the time, and their use in research. The goal was to examine the behavioral characteristics of seafarers. With the cooperation of several young seafarer friends from a merchant marine university in the experiments, the author analyzed collision-avoidance actions using a ship maneuvering simulator. Actions taken under different circumstances were analyzed: in particular, the differences in the crossing angles between the ship being handled and the target. However, scattered results were obtained, with inconsistent behavioral characteristics in each experiment. The actions performed varied even under the same circumstances and with the same seafarer. The author remembers feeling greatly disappointed. His colleagues in the laboratory were engaged in material or towing-tank experiments, wherein they were monitoring the changes in their results due to changes in the experimental conditions. They could derive rules pertaining to physical phenomena from their consistent results. In contrast, the author obtained inconsistent results even under the same conditions, which made deriving any rules difficult. The author learned how challenging it was to research the characteristics of systems in which human beings were a constituent element.

Afterward, the author started research work at Tokyo University of Mercantile Marine and was engaged in the development and use of ship maneuvering simulators. Developments in ship maneuvering simulation were keeping pace with the times. Analog computers in simulators were being replaced by digital computers, and onscreen projections of other ships were being upgraded from silhouettes to computer graphics. Finally, the fifth generation of ship maneuvering simulators was highly realistic and was reviewed favorably by researchers worldwide.

Many personnel connected to the field visited the simulator, and TV and the newspapers reported on it frequently. During a visit from a delegation from the Japan coast guard, the author was quite impressed by what they said: "We had long wanted to examine a cause analysis of the accident that occurred in 1974. It might be possible with this ship maneuvering simulator, since it can quite precisely re-create the actual conditions." Afterward, the author examined the situation of the accident and found that we could recreate the situation by using the ship maneuvering simulator we had developed. Thus, the situations were reproduced, and experiments examining the cause of the accident were initiated. The seafarers participating in the experiments included friends who had participated in the experiments with the first unit I had developed as a student. By this time, they had grown to become experienced seafarers with experience working as captains. This research is what altered my perception of system analysis, including human beings.

The subject was an investigation into the cause of the collision between *No. 10 Yuyo Maru* and the *Pacific Ares* in November 1974. This accident occurred close to the capital city of Tokyo, deep within Tokyo Bay, and resulted in many fatalities and injuries, making it a major focal point of public attention. Furthermore, *No. 10 Yuyo Maru* caught fire, and west winds brought her close to an industrial complex on land. To prevent the

Figure 1 Newspaper article on the circumstances of the collision accident between *No. 10 Yuyo Maru* and the *Pacific Ares*

(Japan Times, dated November 10, 1974)

flames from spreading to the industrial complex, *No. 10 Yuyo Maru* was towed out of the bay and sunk by bombing by the Japan Self-Defense Forces near the bay entrance. Figure 1 is a newspaper article on the immediate aftermath of the accident.

CIRCUMSTANCES OF THE ACCIDENT

Figure 2 illustrates the circumstances of the accident. The LPG tanker *No. 10 Yuyo Maru* proceeded northbound through Nakanose Fairway. The ore carrier *Pacific Ares* had departed from Kisarazu Harbor and was exiting Tokyo Bay by heading toward Uraga Fairway. The reported visibility at the time was 2 miles. A pilot had been on board the *Pacific Ares*, but at the time of the accident, the pilot was already leaving. After the accident, the pilot said that he had informed the *Pacific Ares'* crew of the presence of the northbound *Yuyo Maru* in Nakanose Fairway. However, the *Pacific Ares* had taken no specific collision-avoidance countermeasures prior to the accident. The *Pacific Ares* sank immediately after the collision with all hands lost, so the actual decisions and actions taken onboard remain unknown. At 13:32, the *No. 10 Yuyo Maru*, northbound in Nakanose Fairway, had detected the *Pacific Ares* at the point labeled "P.A. recognized" in Figure 2.

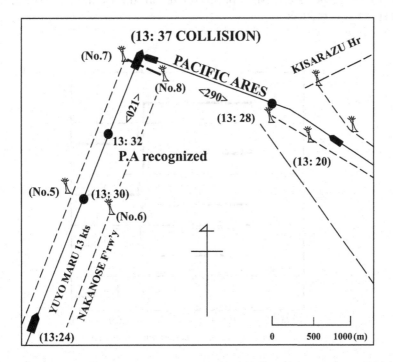

Figure 2 Circumstances of the collision between *No. 10 Yuyo Maru* and the *Pacific Ares*

No. 10 Yuyo Maru continued sailing despite having detected the *Pacific Ares*. She started to decelerate 1400 meters before the collision point, but it was too late.

A marine accident inquiry was convened to investigate the accident. Its judgment was that the *Pacific Ares* was largely at fault and that the actions taken by *No. 10 Yuyo Maru* had been inappropriate as well.

STUDY OF THE ACCIDENT CAUSE INVESTIGATION

Accident recreation and investigation of the cause were started using the ship maneuvering simulator. Ship handling experiments were conducted through a recreation of the *No. 10 Yuyo Maru* accident with the participation of four captains with experience aboard tankers similar to the *No. 10 Yuyo Maru*. The experiments were executed with various changes to the starting time of the maneuvers. An illustrative result is presented in Figure 3. This figure shows the operating status of the main engine under the four captains. The timings for main engine operation match the point at which it was being handled by the captain of *No. 10 Yuyo Maru*. The horizontal (right-side) axis in the figure indicates the collision point (zero). FH, SE, and FA on the vertical axis indicate full ahead, stop engine, and full astern, respectively. No captain ordered an immediate full astern, but the speed was reduced.

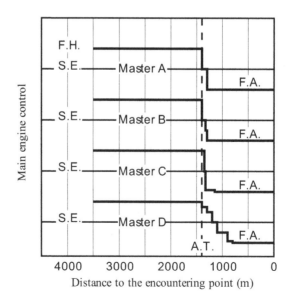

Figure 3 Engine operations under the same conditions by four different captains

Other experiments had shown that handling to change course by steering the rudder was possible, but actual collision-avoidance actions mainly comprised speed reduction.

The maneuvering experiments conducted under various conditions revealed that the four captains handled the ships similarly; the resulting collision-avoidance actions carried a risk of collision or bringing the ship close to the *Pacific Ares*.

The foregoing verification study clarified the following key points:

1. Seafarers with sufficient experience largely take similar actions and make similar decisions under the same circumstances.
2. To prevent a repeat of the accident, the regular behavioral characteristics of seafarers must be considered for countermeasures.

COUNTERMEASURES FOR PREVENTING ACCIDENT RECURRENCE

The *No. 10 Yuyo Maru* accident investigations revealed an important fact: standard seafarers take largely similar actions under the same conditions; this must therefore be a major consideration in countermeasures for preventing accident recurrence, and prevention countermeasures for similar accidents involving human beings should be discussed in light of this theory. Figure 4 is a model of a situation similar to the accident under investigation.

In Figure 4, seafarers demonstrating standard behavioral characteristics (A) in response to an environmental condition (A) execute an action (A), following which the accident occurs. This model reflects the behavioral phenomenon identified by analyzing the *No. 10 Yuyo Maru* accident and illustrates the process through which an accident occurs.

If methods that can avoid accidents in this model could be determined, these would serve as concepts of countermeasures for recurrence prevention. In the following text, the author considers methods that change at least one of the three elements shown in Figure 4. The author describes three cases in the depicted order of the elements and offers recurrence prevention countermeasures.

RECURRENCE PREVENTION COUNTERMEASURE CASE I

The top panel of Figure 5 shows the relationship between the environment and human behavioral characteristics, and the bottom panel shows the countermeasures through which human behavioral characteristics (A) are changed to (B). It shows that when the executed action is changed from (A) to (B), no accident occurs. The question is how to change the behavioral

Figure 4 Model of the circumstances of an accident caused by environmental conditions and the standard behavioral characteristics of seafarers

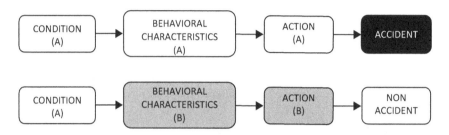

Figure 5 Model of preventing accident recurrence by changing human behavioral characteristics

characteristics from (A) to (B). Because a standard action (A) is taken in response to condition (A), this change would not occur in seafarers taking the standard action.

In the *No. 10 Yuyo Maru* accident, for example, *Yuyo Maru*, proceeding in the fairway, encountered the *Pacific Ares*, that may cross the fairway, then her collision-avoidance action was initiated late because *Yuyo Maru* expected the target to take some collision-avoidance action.

However, since the *Pacific Ares* was not actually intending to cross the fairway, but only to cross outside the end of the fairway, the *Pacific Ares* was under no obligation to give way according to the Act on Preventing Collision at Sea; the ship proceeding in the fairway is thus required to take collision-avoidance action. It is believed that in the case of the *No. 10 Yuyo Maru*, there was confusion regarding the relationship of the ships to the crossing point and who was obliged to take collision-avoidance action. In other words, if the ship proceeding in the fairway had had prior knowledge that the crossing point would not be on the fairway, its actions would have been different. If seafarers were aware of the conditions around a marine traffic vessel in such waters, then the actions taken would be changed from (A) to (B), and no accident would occur. In this case, the pilot is the seafarer who should be aware of the conditions around a marine traffic vessel in such waters.

RECURRENCE PREVENTION COUNTERMEASURE CASE 2

Figure 6 shows the countermeasures in which human behavioral characteristics (A) do not change in response to condition (A), but no accident

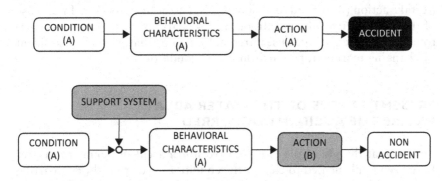

Figure 6 Model of preventing accident recurrence by employing a support system

occurs because of the action nevertheless changing from (A) to (B). The question is how to change the action from (A) to (B). Seafarers showing standard behavioral characteristics (A) need some external support not to perform action (A) in response to condition (A) but to perform action (B) instead. In Figure 6, this external support is represented as "support system". In the case of the *No. 10 Yuyo Maru* accident, such support would be the provision of information by a traffic information center, which means that the ship sailing through Nakanose Fairway is apprised that the ship departing from Kisarazu Harbor will cross outside the fairway exit. With this information, the seafarer can start to decelerate early to prevent a collision. Thus, the action changes from (A) to (B), preventing an accident.

RECURRENCE PREVENTION COUNTERMEASURE CASE 3

Figure 7 shows the countermeasures in which an accident does not occur because the actions of seafarers change from (A) to (B) due to the environmental conditions changing from (A) to (B). As a result of the change in the environmental conditions, the seafarers showing behavioral characteristics

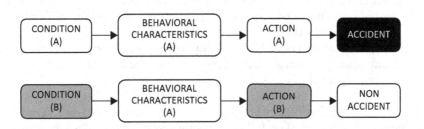

Figure 7 Model of preventing accident recurrence by changing environmental conditions

(A) take action (B) instead of action (A), and no accident occurs. In the case of the *No. 10 Yuyo Maru* collision, this scenario means creating an environment in which standard seafarers take actions that prevent an accident by changing marine traffic environmental conditions.

PRESENT STATUS OF THE WATER AREA WHERE THE ACCIDENT OCCURRED

Presently, at the site of the *No. 10 Yuyo Maru* accident, countermeasures have been implemented to change the conditions such that ships departing from Kisarazu Harbor do not encounter vessels passing through Nakanose Fairway near the north exit of the fairway, as shown in Figure 8.

A new buoy has been set on the extended line of the Nakanose Fairway, and administrative guidance is provided for westbound ships departing Kisarazu Harbor and sailing north of this buoy. This measure has ensured that no ship crossings occur near the fairway exit and has effectively prevented accidents such as the *No. 10 Yuyo Maru* collision from recurring.

The countermeasure depicted in Figure 8 corresponds to case 3 in the preceding discussion. In this case (Figure 9), if the environmental conditions are changed, despite the behavioral characteristics shown by standard seafarers, their action will not be action (A), which results in an accident, but will be changed to action (B), which does not result in an accident.

The effectiveness of the environmental changes and the countermeasures substantiate the theory presented herein, which has helped to strongly advance the theory.

LEARNING FROM THE ANALYSIS OF THE *NO. 10 YUYO MARU* ACCIDENT

The following learnings were verified through analysis of the *No. 10 Yuyo Maru* accident:

1. Standard seafarers handling a ship sailing in a fairway are later than usual in taking action against ships expected to cross the fairway, for example, when there is a risk of collision.
2. Standard seafarers handling a ship sailing on a fairway mainly execute speed reduction maneuvers to avoid the risk of collision with ships expected to cross the fairway.

These learnings highlight the importance of ascertaining the standard behavior of seafarers. The following points from this analysis have been clarified:

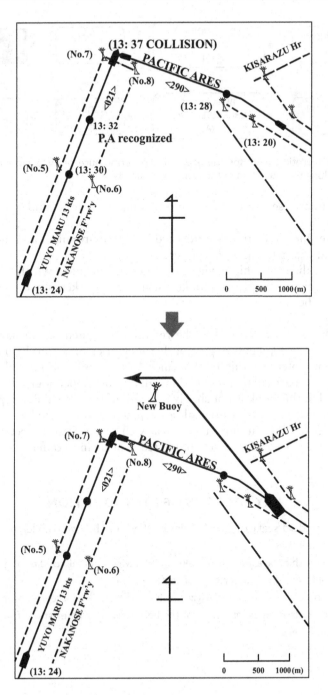

Figure 8 Water conditions at the accident site after countermeasures to change the environmental conditions were implemented

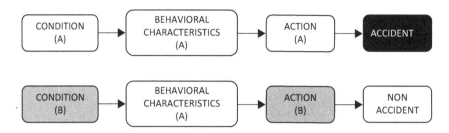

Figure 9 Theoretical basis for the accident countermeasure of installing a new buoy along the extension of the Nakanose Fairway

1. Standard seafarers take largely similar actions under the same conditions.
2. Safe ship navigation is influenced by the environment and the behavioral characteristics of seafarers.
3. To realize safe ship navigation, an environment in which the behavioral characteristics of standard seafarers are taken into consideration must be created.

The author's experiences with the research presented herein have led to substantial developments in previously vague discussions on ship handling. The main content of this book, which begins with the next chapter, is based on the particulars presented herein and developed logically. What is explained in this book is not simply a matter of theoretical development; it is the product of both theoretical development and repeated experiments. The results from the author's research as well as the latest theoretical developments have been successively presented at academic conferences.

KEY FACTORS OF INTRODUCTION

1. Standard seafarers exhibit almost the same behavior under the same conditions.
2. Safe ship navigation is influenced by environmental conditions and the behavioral characteristics of seafarers.
3. To realize safe ship navigation, it is necessary to create an environment that takes into account the behavioral characteristics of standard seafarers.

Part I

Techniques for Ship Handling

Chapter 1 Factors in Achieving Safe Navigation 5
Chapter 2 Analysis of Techniques for Ship Handling 23
Chapter 3 Inadequate Knowledge and Competency Often Observed in
            Inexperienced Seafarer 79
Chapter 4 Significance and Use of Elemental Technique Development 97

Part I

Preface

Ship navigation has a long history and has involved the use of the latest knowledge and techniques in each era in order to meet the demands of maritime transportation. Much has changed from the dugout canoes used to traverse waters in olden times to wooden ships and eventually to the vessels of the steel ship era. Propulsion mechanisms have also changed, from sculls and oars to wind power (i.e., sails) and to the present day propeller-driven propulsion systems. As human civilization advanced, the necessary material and mechanical knowledge regarding ship navigation have been refined. In contemporary times, this progress is driven by the application of scientific methods: analyzing relevant investigative elements and logically elucidating the contribution of these elements individually. These methods are fundamental to modern scientific development, and their own effectiveness has been scientifically proven. In other words, significant developments in the physical and structural study of ships have been achieved through the application of modern scientific techniques. Thus, such studies have led to the development of individual studies and the formation of relevant systems.

The navigation techniques used by seafarers are indispensable in any type of sailing, and navigation techniques designed for sailing over rivers and lakes have been modified and expanded to suit ocean navigation. Furthermore, these techniques, which must adapt to the changing conditions in the navigational areas, are becoming increasingly sophisticated as a result of numerous studies on navigation planning, determining ship position, and maneuvering vessels in order to realize the navigational objectives. Considering the state of contemporary ship navigation, what techniques might be considered "necessary"? Typical examples are those listed in the 1978 International Convention on Standards of Training, Certification and Watchkeeping for Seafarers, which was adopted by the International Maritime Organization, an agency of the United Nations. These examples are partially referenced in this book. However, from a modern scientific perspective, they cannot be always regarded as the product of logical analysis. Since seafarers require clear definitions and functional analysis of necessary techniques in order to ensure and maintain safe navigation, the necessary conditions must be clarified, which would also promote the further development of techniques.

Accordingly, the aim of Part I is to organize navigation techniques that have been assumed impossible to systematize in terms of knowledge or science. It is not intended to explain the practical skills of navigation in detail, or all the knowledge necessary for executing those skills.

The techniques for achieving safe navigation are typical of techniques for the effective operation and management of huge systems in modern society. Moreover, appropriately recognizing such advanced techniques is indispensable to ensure the safety of those systems. Hence, it is hoped that this book will help promote a greater understanding of the high degree of competency which operating engineers must show to manage huge systems efficiently and safely, and also the importance of taking this competency into careful consideration.

Chapter 1

Factors in Achieving Safe Navigation

1.1 DIFFICULTY OF THE NAVIGATIONAL ENVIRONMENT

Various competencies are essential to ensure safe ship navigation. The techniques needed in different areas of operation are categorized on the basis of the functions essential for safe navigation.

In this context, technique is "the particular way in which we achieve objectives", competency means "the ability to execute the techniques", and function means "the role for achieving purpose".

The techniques needed for a given navigational situation are selected from among the various available techniques and are then implemented. These techniques are considered in detail in Chapter 2. In this chapter, among the necessary techniques for safe navigation, let us consider position fixing. Position fixing is broad in meaning and can be defined as any action involving the estimation of a ship's geographic position. The level of technique required for position fixing is dependent on the required estimation precision: when navigating on open waters, a margin of error of approximately 3 miles is acceptable, but the acceptable error is very small when entering a narrow channel.

The horizontal line in Figure 1.1.1 represents the difficulty of the navigational environment, where the left end of the line is the origin, and the situation becomes increasingly difficult as one moves to the right, farther away from the origin. Let us consider the factors that determine environmental difficulty. The environment at point "a" has higher navigational difficulty than the environment at point "b". The factors determining navigational difficulty in detail are dealt with in Section 1.2. For example, one is a situation in which the conditions at point "b" are an open sea 50 miles from the shore with calm weather. Another situation is an open sea area, point "a", within 10 miles of the shore, with heavy traffic and with weather conditions of winds exceeding 15 meters per second; these conditions are more navigationally difficult than those at point "b".

As may already be apparent to the reader, this situation presents the following questions: What is meant by "difficulty", and what might the units

Figure I.I.I Difficulty in navigational environments

be for indicating difficulty on the line? What are the locations of points "a" and "b" on the line? What is the distance between them? These problems have been academically discussed for a long time. This book will provide the reader with useful insights into these issues.

Although these concerns may be abstract, we can proceed with our discussion if you agree that the difficulty[1] of achieving safe navigation differs depending on the water area and navigation conditions.

---

[1] Note: The word "difficulty" inherently expresses how hard it is to achieve an objective. The elements that determine difficulty are governed by the relationship between the given conditions under which an objective is to be achieved and the ability to achieve that objective. When discussing navigational difficulty in common water areas, conditions in the navigable areas, traffic volume, and weather and sea state are often considered as factors. However, note that difficulty is usually estimated with the tacit assumption that seafarers have specific abilities to achieve certain goals; in other words, standard competency.

1.2 FACTORS AFFECTING NAVIGATIONAL DIFFICULTY

In the previous section, the influence of navigational conditions on achieving safe navigation was discussed. In other words, change in the difficulty of continuously maintaining safe navigation was considered. It has become clear that such difficulty changes according to relevant factors. Following these views, our discussion proceeds to the next issue: the navigational environment is never constant, even in the same water area.

Sometimes, the number of maritime traffic vessels increases compared with the average. In addition, sometimes, visibility is restricted by dense fog. The navigational difficulty in oceanic areas shown in Figure I.1.1 may be considered average and determinable under certain fixed conditions. Thus, if we consider that water areas indicated by the average difficulty at point "b" sometimes present different difficulties then the difficulty may be changeable caused by varying conditions. Such changes in conditions may be presented in terms of the concept of the probability of an event occurring. Figure I.1.2 modifies Figure I.1.1 by adding the event probabilities of changes in the conditions occurring, with the conditions of average difficulty as the centers.

Conditions with the most frequently appearing difficulty in each area are indicated by point "a" and point "b". Thus, although lower in frequency than the average conditions, more difficult conditions may appear. On the other hand, there are also conditions under which the difficulty could

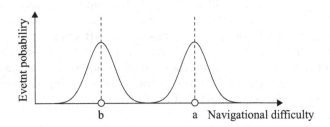

Figure I.1.2 Difficulty of navigational environment and event probability

Table I.1.1 Factors determining navigational difficulty

1 Ship maneuverability
2 Geographic and water conditions being navigated
3 Weather and sea state
4 Marine traffic conditions
5 Rules of navigation
6 Onboard handling support systems
7 Onshore navigation support systems

change to lower than average. Figure I.1.2 shows that the possibility of the occurrence of conditions that are even further removed from the average difficulty is even smaller.

We next consider in detail the factors that determine navigational difficulty, which are listed in Table I.1.1.

1) Ship maneuverability

The turning radius, stopping distance, and other maneuvering characteristics of the ship being handled are directly connected to its difficulty in maneuvering, especially in water areas that are narrow or have congested traffic. In addition, very large ships such as very large crude carriers (VLCCs) have poor speed deceleration performance and therefore, require a large time allowance in order to avoid collisions by adjusting speed. In contrast to small ships, which can quickly adjust speed to avoid imminent risks, large ships face great difficulty in achieving safe navigation under such conditions.

In addition, changes in the difficulty of controlling a ship because of the effects of external forces need to be considered. Car carriers and container ships have a large surface area above the water line that is subject to wind pressure, and turning motion and lateral movement due to wind pressure influence the difficulty of movement control, but this is not always defined as maneuverability. The form of a ship's upper structures is also considered a factor that affects ship safety.

2) Geographic features and water conditions being navigated

The expanse and form of the water area being navigated are factors that affect the difficulty of handling ship movement. In recent years in particular, waterway conditions, specifically curving, width, and water depth, have been major factors affecting the difficulty of navigating large vessels. Furthermore, the conditions of water areas being navigated are closely related to the conditions of the water areas available to avoid risks. The constraints on realizing safe navigation are the factors altering navigational difficulty.

3) Weather and sea state

Restricted visibility due to fog, rainfall, and snowfall hinders lookout duties, which are based on vision and are bases of safe navigation; these obstacles make ship handling and ensuring safety more difficult for seafarers. In addition, ocean currents and large waves in narrow waters greatly magnify the difficulty of maintaining safe navigation. Moreover, as explained earlier, wind and other external forces affect the degree of control over ship movements and thus must also be considered elements affecting navigational difficulty.

4) Marine traffic conditions (types and volume of traffic vessels)

Many traffic ships navigating in a limited water area increases the difficulty of lookout duties. Furthermore, complicated encounters increase collision risks and navigational difficulty.

5) Traffic regulations

Traffic regulations were enacted to ensure safety in waters frequented by an increasing number of ships and by ships of larger sizes, whereas before, Freedom of Navigation was allowed. The purpose of these rules was to regulate previously random ship traffic flow and increase safety through navigation based on fixed rules. Regulations corresponding to conditions in navigational areas contribute to the degree of safety and decreasing navigational difficulty. Examples of how the enactment of rules of navigation contributed to improving safety in ship navigation will be presented later.

6) Onboard handling support systems

Various nautical instruments, such as RADAR/Automatic Radar Plotting Aids (ARPA), Electronic Chart Display Information Systems (ECDIS), and Automatic Identification Systems (AIS), are installed on a ship's bridge in order to lessen the workload on seafarers and to increase navigation safety.

7) Onshore navigation support systems

Information on traffic vessels, weather in the planned areas of navigation, and fishing boats in operation within a specific area can be obtained from onshore support systems. This information is useful in estimating uncertain future conditions in order to lessen navigational difficulty.

Examples of Rules of Navigation Contributing to Ship Navigation Safety

This section explains how the rules of navigation described under item 5 (i.e., traffic regulations) in the previous section affect the difficulty of navigational environments.

Figure I.1.3 illustrates traffic management over the arrival time of ships in a water area under condition "a", where the average difficulty is μ_E. Assume that this management is intended to ensure that the intervals between ship arrivals are at least a specified minimum, which avoids the concentration of too many ships within a short period of arrival.

This management creates a situation in which there is no crowding of arriving ships, which consequently avoids very difficult navigational conditions that entail navigating through a congested water area. However, because this approach does not affect the total number of ships in the area, the probability of short and long intervals between ship arrivals becomes small; that is, such management prevents crowding in the short term, reduces the duration of inactive periods (i.e., periods with few ships), and increases the probability of average traffic volume, as reflected by the large change in event probability of conditions with average difficulty in Figure I.1.3.

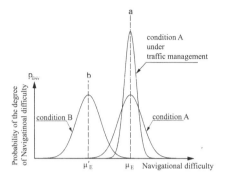

Figure I.1.3 Effect of traffic management regulating arrival time intervals on navigation difficulty

In general, the objective of traffic management is to reduce the occurrence of risky situations. Management reduces the crowding of ships on arrival and increases the probability of average conditions. Therefore, the foregoing method of representation can be used to explain changes in the navigational environment due to management, and the diagram in Figure I.1.3 is thus a valid method of representing the difficulty of navigational conditions.

Competencies of Seafarers to Overcome Environmental Difficulty

As environmental difficulty increases, advanced levels of competencies are required to ensure safety. Therefore, to maintain safety under particular conditions, the competency level must correspond to the expected environmental difficulty. Narrow navigational areas with a high degree of difficulty demand highly precise position-fixing competencies with little room for error. Therefore, the competency level needed for safe navigation in water areas with a highly difficult navigational environment must be correspondingly high. Accordingly, it can be concluded that *the difficulty of the navigational environment determines the necessary competency level.*

In Figure I.1.4, the environmental difficulty represented on the horizontal axis in Figure I.1.2 is substituted with *the competency level required for an*

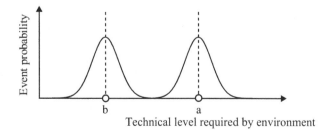

Figure I.1.4 Technical level as required by the environment

environment, and the vertical axis represents the probability of conditions changing (i.e., changes in difficulty). Assume that points "a" and "b" indicate the average conditions. Then, the probability of a change in the conditions can be depicted as a probability distribution curve with centers at points "a" and "b". Of the factors determining environmental difficulty listed in Table I.1.1, factors 2), 5), 6), and 7) can be considered as usually fixed conditions, whereas factors 3) and 4) are conditions that may change at any time. When the conditions and consequently the difficulty change, having seafarers with a high competency level ensures safe navigation even under the increased difficulty.

KEY FACTORS OF SECTION 1.2: FACTORS AFFECTING NAVIGATIONAL DIFFICULTY

Navigational difficulty is influenced by the following factors:

1) Ship maneuverability
2) Navigational geographic conditions, such as form of navigable waters and water depth
3) Weather and sea state
4) Marine traffic
5) Rules of navigation
6) Onboard handling support system
7) Onshore navigation support system

1.3 SHIP HANDLING COMPETENCY OF SEAFARERS

Ensuring safe navigation can be considered a matter of whether the seafarer's competency fulfills the technique require for the navigation environment. The line in Figure I.1.5 indicates the achievable competency of seafarers, where the competency increases as one moves farther to the right along the line. Thus, in the figure, the competency of seafarer "b" is higher than that of seafarer "a".

Demonstrable seafarer competency varies with each of the following factors:

1) Seafaring license rank held by seafarers
2) Actual navigational experience at sea
3) Fatigue (related to length of working hours and time elapsed standing watch)
4) Stress (related to such conditions as seafarer awareness)

The vertical axis in Figure I.1.6 indicates the event probability for achievable competency of seafarers. Seafarers having the same seafaring license rank and similar experience will not necessarily have the same competency. Generally, competency will vary depending on the individual seafarer. In the figure, this phenomenon, that demonstrated seafarer competency is not always fixed, is represented as a curve showing event probability. Moreover, the achievable competency of even an individual seafarer is not always fixed; changes in an individual's achievable competency can be thought of as changes in the activity of that person.

Figure I.1.5 Representation of seafarer competency

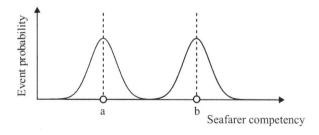

Figure I.1.6 Seafarer competency and event probability

Table I.1.2 Five phases of awareness

Phase	Awareness
0	Unconscious, sleep
I	Careless, drunken
II	Normal, relaxed
III	Normal, active
IV	Panic

Table I.1.2 lists awareness levels as defined in the field of cerebral physiology: Phase 0 indicates a state of being completely unaware, such as during sleep; Phase I indicates decreased awareness, such as when consuming alcohol; and Phase II is a relatively normal state in which one is relaxed and can demonstrate normal awareness. Phase III indicates a state of very high activity in which the brain is alert. People in such a state are said to demonstrate greater than normal capabilities: they can process information quickly and with a high degree of accuracy. If one could constantly achieve the Phase III state, one would always be able to respond to difficult situations. However, according to cerebral physiology literature, maintaining a high degree of attention for prolonged periods is difficult, and Phase III can at best be maintained for 5–10 minutes. Phase IV indicates a state of panic in which a person is too tense and cannot make systematic decisions. A stress-induced panicky state is something to watch out for.

Table I.1.2 suggests that human information processing capabilities change depending on the situation, which should be considered while discussing the achievable competency of seafarers. Even for an individual seafarer, the achievable abilities must be considered not as being constant but as something that changes depending on various factors. The average competency shown in Figure I.1.6 is the ability demonstrated in Phase II, meaning that the state may degrade to Phase 0 or I depending on the circumstances. In addition, this means that at certain times, greater than average competency, as in Phase III, may be demonstrated. The concept outlined in Figure I.1.6 is thus necessary to cover all possible changes in achievable individual competency induced by physical and mental conditions.

Let us consider changes in the competency of a seafarer. In Figure I.1.7, seafarer "a" presenting μ_H on average is in Phase II. Assume that the state of this seafarer is approaching Phase 1 because of factors such as fatigue. In the figure, this state is labeled β. The state in which the seafarer can demonstrate higher than normal level of competency, "α" state, is a competency level equivalent to Phase III. It is labeled α.

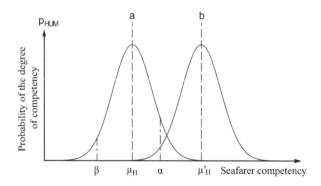

Figure 1.1.7 Changes in the degree of seafarer competency because of changes in awareness

KEY FACTORS OF SECTION 1.3: SHIP HANDLING COMPETENCY OF SEAFARERS

The ship handling competency of seafarers is influenced by the following factors:

1) Seafaring qualifications of seafarers
2) Actual onboard navigational experience
3) Degree of seafarer fatigue
4) Degree of seafarer stress

1.4 CONDITIONS NECESSARY FOR SAFE NAVIGATION

Section 1.2 explained how navigational difficulty is determined by the environmental conditions in a specific water area. Therefore, for safe navigation, seafarer competency must be adequate for the water area difficulty. That is, the necessary seafarer competency is the competency required for a given environment. We next discuss how to balance these two factors in order *to realize safe ship navigation*. Figure I.1.8 illustrates the relationship between these two factors, where the horizontal axis shows the competency required for the environment, and the vertical axis shows the achievable competency of seafarers. The straight 45° inclined line in the figure indicates the points at which both factors have the same value. In other words, if the state indicated by the straight line can be ensured, then the competency required for the environment and the achievable competency by seafarers are the same, which means that safe ship navigation can be realized. The region above the straight line shows where seafarers can realize higher competency than that required for the environment and consequently, where safe navigation is achievable, whereas the region under the straight line indicates where seafarers cannot realize the competency required for the environment and consequently, where navigation is dangerous. Hence, this 45° line can be considered as indicating the boundary between safe and dangerous conditions.

Figure I.1.8 shows the basic concept underlying the determination of safe ship navigation: safety in ship navigation can be achieved when seafarers have sufficient competency for a given environmental condition. If seafarers have the standard competency but still cannot manage safe navigation, then the navigational environment must be improved.

Figure I.1.9 presents the variations in seafarer state and environment, described in Sections 1.2 and 1.3, as well as the relationship between the competency required of seafarers when the elements determining environmental conditions change and the achievable competency of seafarers.

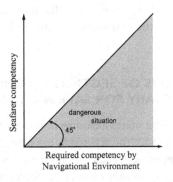

Figure I.1.8 Conditions necessary for safe navigation according to the relationship between competency required for the navigational environment and seafarer competency

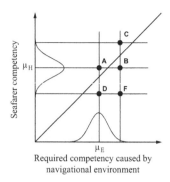

Required competency caused by
navigational environment

Figure I.1.9 Changes in ship navigation safety because of changes in environment and variations in seafarer competency

In Figure I.1.9, μ_E is the mean competency required for the environment, and μ_H is the mean seafarer competency. When both factors are average, the resulting condition, labeled situation "A", is a state above the 45° line, which means that safe navigation is achievable. By contrast, when the environmental conditions worsen and navigational difficulty increases, seafarers maintain the mean competency, and the state shifts rightward to situation "B". In this situation, the competency that can be realized by seafarers is lower than the required competency, thus indicating a change to a dangerous situation. In response, if seafarers increase their concentration and become capable of a higher degree of information processing (i.e., if seafarer competency increases), then the situation shifts to "C", which is above the 45° line, meaning that safe navigation is possible. Furthermore, if seafarers exhibit lower competency as a result of fatigue, for example, even if the environmental conditions are average, the situation shifts toward "D", which is a state under the 45° line, where safe navigation cannot be achievable. If seafarers show even lower competency, and the environmental conditions worsen further, the situation becomes even more dangerous (situation "F"). The further the vertical distance of a state from the 45° line, the more dangerous the situation, and the higher the risk of accidents.

KEY FACTORS OF SECTION I.4: CONDITIONS NECESSARY FOR SAFE NAVIGATION

1. Safe navigation necessitates a balance of "the competency required for the environment" and "the achievable competency of seafarers".
2. When "the competency required for the environment" is higher than "the achievable competency of seafarers", safe navigation is difficult to realize.

1.5 TECHNIQUES NECESSARY FOR SAFE NAVIGATION

As described in the previous section, ships encounter various maneuvering situations between departing a port and arriving at the intended destination. In response, seafarers must execute the necessary techniques to handle the ships and achieve the intended navigation. In doing so, the first objective of the seafarers is to achieve safe navigation without accidents. The major accidents a ship may encounter are collisions with other traffic vessels and grounding accidents in shallow waters. In addition, capsizing accidents may rarely occur during storms and other rough weather.

In this book, we consider the techniques for realizing safe navigation under the regularly encountered weather and sea conditions.

Consistent exhibition of seafaring competency is required to achieve safe navigation. This is not only mandated by national regulations but is also dictated by the international treaty of the International Maritime Organization (IMO), an agency of the United Nations: specifically, the 1978 International Convention on Standards of Training, Certification and Watchkeeping for Seafarers (STCW).

The stipulations of the STCW are broadly categorized in Figure I.1.10. Among these, "competence" is defined as the ability to execute the techniques necessary to realize safe navigation. Therefore, an understanding of *the necessary techniques* for performing safe navigation is vital.

How to evaluate competence is a natural question at this point. What are the actions through which the necessary competency can be assessed? Depending on what competency is being assessed and how the evaluation criteria are expressed, the results of the evaluations might not be uniform, making it difficult to ensure consistent criteria for seafarer competency.

Figure I.1.11 partially details how the criteria for evaluating the necessary competency and knowledge of competency are written. In the STCW, "competence" is treated as a basis of ship navigation. In this book, "techniques" and "competency" are treated as distinct entities. *"Techniques"* are defined as the sum of functions performed to achieve safe navigation, whereas *"Competency"* is defined as the ability to perform the necessary

Knowledge, understanding, and proficiency:
        Basic knowledge of, understanding of, and proficiency in competency.
Methods for demonstrating competence:
        Methods for assessing competency.
Criteria for evaluating competence:
        Criteria for assessing competency.

Figure I.1.10 Competency stipulations by the STCW

Table A-II/2 in Chapter II of the STCW code

Competence: Positioning

- Knowledge, understanding, and proficiency: Position determination under all conditions.
- Criteria for evaluating competence: The primary method chosen for fixing the ship's position is the most appropriate for the prevailing circumstances and conditions. The obtained fix is within accepted accuracy levels, and the accuracy of the resulting fix is properly assessed.

Figure I.1.11 Competency criteria for position fixing in the STCW

techniques. In other words, techniques are clearly defined for safe navigation. However, the level of achievement of the functions required for techniques (i.e., competency) varies among seafarers. Consequently, seafarer abilities are evaluated in terms of competency.

Figure I.1.11 shows only those parts relevant to position fixing, but similar descriptions are available for other competencies, such as maneuvering. The techniques for position fixing listed in the figure must be performed to exhibit position-fixing competency. Let us study these items listed in the order they appear.

First, "knowledge, understanding, and proficiency" refers to the knowledge and understanding of and proficiency in "ship position determination *under all conditions*", which is the *technique needed to fulfill the necessary functions*.

"Criteria for evaluating competence" is as follows: "The primary method chosen for fixing the ship's position is *the most appropriate for the prevailing circumstances and conditions*. The obtained fix is *within accepted accuracy levels*, and the accuracy of the resulting fix is *appropriately assessed*".

In other words, the following competency must be demonstrated:

1) The primary method for position fixing is the one most appropriate for the conditions of the water area being navigated and the circumstances of the installed instruments.
2) Thus, for the obtained ship position, the precision is appropriate for the circumstances of the ocean area being navigated.
3) To evaluate the accuracy of the positioning results, the method used must be considered, and the ability to evaluate the factors affecting positioning is vital.

Seafarers should focus on the italicized sections of "criteria for evaluating competence". A commonality between all the italicized sections is that the expressions are vague. The STCW does contain some supplementary explanations in addition to those presented in the figures, but they are equally vague. Although these are general descriptions for international regulations, they are also intended to serve as standards; therefore, the possibility of multiple interpretations should be avoided. Otherwise, these vague expressions may result in different interpretations by educators and trainers, which may cause differences in the education and training itself. This may then result in differences in the criteria for assessing competency. Seafarers thus educated by different educators and trainers would therefore not have the same competency, meaning that seafarers of different competency levels would possess the same seafaring qualifications.

The objective of the STCW is for seafarers to possess standardized, uniform, and consistent levels of competency. To this end, it is indispensable that the specific details of the techniques, as well as those of competencies corresponding to various seafarer competency levels, are explicitly stated.

In the 1990s, the concept of a "functional approach" to ship navigation was internally proposed within the IMO, with the objective being *the categorization and evaluation of the techniques necessary for safe navigation*; that is, clarifying the functions and categorizing how they should be fulfilled by standard seafarers. However, this idea remained a proposal, as further studies on it were abandoned without any analysis.

In Japan, the results from a study of seafaring techniques, "Development of Elemental Techniques for Ship Handling", were presented at the Japan Institute of Navigation in 1997. This study analyzed the onboard functions of seafarers and explained how different actions contribute to navigation safety. Several hundred functions were categorized according to their objectives.

For example, measuring the direction toward a target is often done onboard by using a bearing compass. However, the function that this activity serves differs depending on the objective of this measurement: the direction of a target on land can be measured for estimating the ship position, whereas the direction of other traffic vessels is measured to estimate the collision risk. Thus, from the perspective of ensuring safe ship navigation, the *objective of an action* is more important than the actual action. Remember that onboard actions are executed only for achieving the necessary functions. In other words, actions that do not achieve the necessary functions are comparable with those that do not contribute to safe navigation. Thus, *techniques necessary for safe navigation cannot be analyzed solely by analyzing the onboard actions of seafarers without considering their purpose.*

In the aforementioned Japanese research "Development of Elemental Techniques for Ship Handling", various actions as well as their intended functions were analyzed, and the various actions that must be performed by seafarers in order to realize the functions necessary for safe ship navigation were identified. These functions can be considered the techniques necessary for realizing safe navigation. The researchers classified these essential techniques into nine technical system categories. This analysis of "What functions must be achieved for safety?" corresponds exactly with the functional approach proposed within the IMO and is a perspective different from the past organization of necessary competencies as specified in the STCW.

Their analytical research clarified the techniques necessary for safe navigation, and the degree to which the techniques are achieved was clearly defined as an indication of seafarer abilities; that is, their competency. In the following section, the identified necessary functions for each technique are defined.

KEY FACTORS OF SECTION 1.5: TECHNIQUES NECESSARY FOR SAFE NAVIGATION

In this section, techniques and competency pertaining to ship handling are defined as follows:

- Techniques are functions necessary to achieve safe navigation.
- Competency is the ability to perform the necessary techniques.

1.6 APPLICABLE NAVIGATIONAL SITUATIONS

Ships face various navigational situations, such as those listed herein, between departing a port and arriving at the intended destination.

1) Docking and undocking maneuvers
2) Harbor maneuvers
3) Restricted water navigation (e.g., fairways and rivers)
4) Coastal navigation
5) Ocean navigation

The navigation techniques discussed in this book are applicable to all the listed situations and are indispensable basic techniques for safe navigation. Here, we consider the prevention of accidents common in these navigational situations.

The main requirement for realizing safe navigation is the prevention of grounding and collision accidents. The probability of these accidents occurring differs depending on the actual navigation conditions. This probability is decided by the difficulty of navigation. For example, there is little danger of grounding in ocean and coastal zones with wide navigable waters, as few other ships would be encountered in these zones. In these situations, a low level of precision and frequency of position fixing and monitoring of other vessels is allowed. However, in coastal areas with restricted navigable waters and high ship traffic, restricted waters, and situations involving harbor maneuvers, highly precise position fixing is often required. In addition, other ships are frequently encountered, which increases the probability of collision accidents; thus, highly precise and frequent lookouts for vessels as well as awareness of present ship status and estimation of potential future collisions are required.

In Chapter 2, by developing the elemental techniques, the techniques necessary to realize safe navigation in each of these navigational conditions will be explained in detail.

Chapter 2

Analysis of Techniques for Ship Handling

The International Convention on Standards of Training, Certification and Watchkeeping for Seafarers (STCW) code of the International Maritime Organization (IMO) describes the functions that seafarers must execute in order to realize safe ship navigation. However, the description used to explain these functions implies that the same techniques are necessary in different navigational situations. Thus, there are overlaps in the techniques required of seafarers in different navigational situations. However, as explained earlier, the descriptions of the vital techniques are unclear.

Clarifying the specific details of the necessary handling techniques is essential for improving the quality and effectiveness of the related training. To this end, investigative commissions of specialists on handling techniques have been convened. To categorize the navigation techniques and clarify the details of individual techniques, ship handling specialists from maritime universities, maritime technical colleges and institutes of sea training, shipping companies, and other seafaring educational organizations have gathered together and engaged in intense discussions. Techniques for ship handling have advanced and been refined in a complex manner over a long period of time. Over time, the necessity for some navigation techniques has diminished, whereas that of others has increased. Although the importance of ship handling techniques is well recognized at present, there has been no overall discussion or organization of techniques in order to identify the overall structure.

Why is it necessary to analyze, organize, and clarify navigation techniques and to identify the overall structure? What are the benefits of identifying any factors in general? These questions are addressed in detail in later chapters, so merely the importance of clarifying techniques is illustrated here.

1) The techniques necessary for maintaining safety should be described clearly and concretely. The necessary techniques vary with the environmental conditions, which in turn change with

each ship maneuvering situation. Achieving the necessary techniques required for the environmental conditions will ensure safe navigation.

2) The relationship between the techniques for maintaining safety and the competency of the seafarer is clarified. Competency can be defined as the level to which a technique can be achieved. Seafarers must master the necessary competencies, so the details of competency training should be clarified in order to facilitate seafarer training. In addition, it would become possible to accurately and uniformly evaluate the competency possessed by seafarers.

3) The human limits of seafarers in performing techniques should be identified. An environment must be created in which safe navigation is possible within the achievable technical competency of standard seafarers. In addition, the safety of any navigational environment can be evaluated using standard seafarer competency as the criterion.

4) Support systems that can expand the achievable competency of seafarers should be conceived.

5) By evaluating marine accidents on the basis of standard seafarer competency, it becomes possible to identify whether the seafarer or the navigational environment is the cause of an accident; accordingly, effective accident prevention measures can be proposed.

As stated earlier, the techniques performed by seafarers for safe navigation have been systematically organized into nine categories, which are hereafter referred to as the *nine elemental techniques*. In this section, each of these techniques is introduced. Table I.2.1 lists the nine techniques and their abbreviated expression.

Table I.2.1 List of the nine elemental techniques

Technique of passage planning	Planning
Technique of lookout	Lookout
Technique of position fixing	Position fixing
Technique of maneuvering	Maneuvering
Technique of observing rules of navigation and other laws and regulations	Observing laws and regulations
Technique of communication	Communication
Technique of instrument operation	Instrument operation
Technique of handling emergencies	Emergency treatment
Technique of management: managing techniques and people	Management

These nine techniques should be applied in combination to avoid collisions, prevent grounding, and otherwise realize safe navigation during actual ship handling. Accordingly, these techniques can be regarded as the basic elements for achieving safe navigation, and thus are termed *elemental techniques*, and the act of organizing the techniques actually performed in the maneuvering situation into elemental techniques is called *elemental technique development*.

Figure I.2.1 shows the conditions essential for seafarers to realize safe navigation. The octagonal column represents the circumstances essential for seafarers to execute the necessary actions. The lowest layer of the column shows that physical health is a necessary condition, meaning that illness, sleep deprivation, and alcohol intoxication impede the achievement of the functions in the layer above. The subsequent layer shows that a mental health condition free of anxiety and worry is essential. The next layer indicates the necessity of basic knowledge, including the ability to accurately grasp situations and logically analyze and solve them. The following layer indicates the necessity of engineering knowledge, including knowledge of mathematics, physics, electricity, and electronics. The topmost layer indicates the position of the nine elemental techniques for ship handling. These techniques are explained comprehensively in the following sections; here, let us consider the conditions necessary for seafaring, which is represented on the topmost layer. This figure conveys that seafaring alone cannot achieve its own objectives; knowledge of engineering is essential, as is the ability to appropriately analyze situations. Furthermore, the figure illustrates that knowledge and analysis cannot be optimally utilized

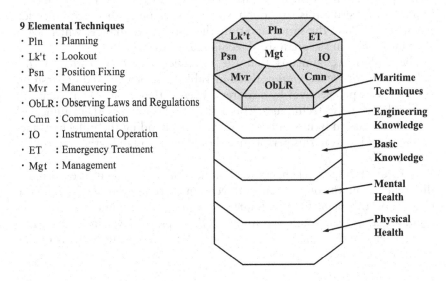

9 Elemental Techniques
- Pln   : Planning
- Lk't   : Lookout
- Psn   : Position Fixing
- Mvr   : Maneuvering
- ObLR : Observing Laws and Regulations
- Cmn   : Communication
- IO   : Instrumental Operation
- ET   : Emergency Treatment
- Mgt   : Management

Figure I.2.1 Conditions essential for seafarers to realize safe navigation

without physical and mental health. Inadequate lookout because of fatigue or delayed recognition of a dangerous situation because of the lookout being pre-occupied has resulted in accidents. Occasionally, errors are made in determining the mutual headings or the situation in which two ships encounter each other because the RADAR information indicating the relative movement of other ships was misinterpreted. Thus, in addition to seafaring itself, the lower layers indicated in Figure I.2.1 crucially influence seafaring. However, occasionally, the depicted structure is forgotten, and problems are consequently misidentified when discussing the causes of accidents.

In this chapter, the navigation techniques shown in the topmost layer of the structure are discussed.

Here, each of the elemental techniques is explained, and their *definitions*, the *functions achieved by performing these techniques*, and the *external factors affecting the achievement of these functions* are discussed (see Table I.2.2). In the STCW, the functions achieved by these techniques are described based on each of the maneuvering situations. Even if the maneuvering situations are different, the functions that must be achieved are described similarly. On the contrary, by applying nine elemental developments, we can understand that the techniques it is necessary to achieve can be listed for different situations based on the necessary functions. In addition, each technique covers several functions, meaning that this provides an opportunity to recognize incomplete understanding or performance of the technique, including the functions, when a mariner cannot achieve them.

Table I.2.2 summarizes the factors influencing the achievement of these techniques. Herein lie the problems in the way the STCW format addresses maneuvering situations: although situations change, the recommended techniques remain the same, and only the conditions change, which leads to inadequacies in the analysis of techniques in the STCW.

The author recommends that the *influencing factors related to the achievement of the techniques* listed in Table I.2.1 be read and understood from various perspectives by seafarers. Even if they believe that they can fully achieve these techniques, could they really execute them under all related changes in influential external factors? Reading the influential factors in the table from this perspective should be useful in further improving competency.

In Figure I.2.1, the elemental techniques for management are arranged differently from the other elemental techniques; this is elaborated on in the chapter on the elemental techniques for management. Figure I.2.2 shows the navigation techniques pertaining to the topmost layer in Figure I.2.1. The key elements of these techniques, which are the primary focus of this book, are indicated.

Table I.2.2 Definitions, functions, and factors influencing the achievement of functions of the nine elemental techniques

Elemental technique	Details
1. Planning (Definition)	Techniques of collecting information on the navigational environment, creating passage plans, and creating plans to execute these plans.
(Main functions)	(1) Understand the information necessary to create plans. (2) Understand the methods of using the necessary information. (3) Apply planning information to the actual plans. (4) During navigation, make changes to plans if the situations are different from those predicted during the initial planning.
(Influencing factors)	(1) Traffic rules and regulations (2) Quality and quantity of effective information for navigation (e.g., presence of the recommended fairway) (3) Quality and quantity of obtainable basic information (information on weather, seas, geographical features, waters, and navigation) (4) Navigation area (ocean navigation, coastal navigation, narrow waterway navigation, fairway navigation, harbor navigation, river navigation) (5) Purpose of navigation (navigation at sea, dropping anchor, docking)
2. Lookout (Definition)	Techniques of detecting stationary targets and moving targets; identifying them; estimating the type, distance, direction, moving speed, and moving direction of targets; and predicting future risks.
(Main functions)	(1) Identify the present situation (types of ships encountered, position and movement of target ships [i.e., course, speed]). (2) Predict the future situation (movement of target ships [i.e., future position, course, speed], changes in the movement of targets, estimated risks to own ship [CPA, TCPA, BCR]).
(Influencing factors)	(1) Navigational instruments (compass, RADAR, RADAR/ARPA, AIS, VTIS information) (2) Visibility (3) Volume and flow characteristics of marine traffic vessels (4) Navigational conditions (in oceans and fairways) and traffic laws
3. Position fixing (Definition)	Techniques of estimating own-ship position by selecting optimal objects visually and by using navigational instruments. Techniques of estimating factors affecting own-ship movements and their magnitude.
(Main functions)	(1) Select methods of collecting information for position fixing (select measuring instruments; select objects for position fixing). (2) Estimate own-ship position (achieve required accuracy and frequency). (3) Estimate own-ship movement status (estimate direction of movement, speed of movement, rate of turn, wind, and tide).

(Continued)

Table I.2.2 (Continued) Definitions, functions, and factors influencing the achievement of functions of the nine elemental techniques

Elemental technique	Details
(Influencing factors)	(1) Types of instruments available for position fixing (compass, RADAR, GPS, echo sounder) (2) Condition of navigational environment (navigation area, available objects for positioning) (3) Visibility (4) Disturbing elements (wind and tide, electromagnetic wave propagation conditions)
4. Maneuvering (Definition)	Technique of controlling the course, speed, and position of the ship through such actions as control of rudder and the main engine.
(Main functions)	(1) Measure movement. (2) Select and choose ship operation equipment (e.g., rudder, main engine, side thruster, tugboat, anchor, and mooring line). (3) Determine operational power (under conditions of both single and simultaneous multiple-device operation).
(Influencing factors)	(1) Maneuvering objectives (e.g., keeping course, controlling own-ship position, controlling speed, and docking) (2) Available control devices (rudder, main engine, side thruster, tugboat, anchor, and mooring line) (3) External forces (wind, tide, water depth, ship–ship interaction, bank effects)
5. Observing laws and regulations (Definition)	Techniques of navigating according to the International Regulations for Preventing Collisions at Sea, Maritime Traffic Safety Act, Act on Harbor Regulations, and other regulations.
(Main functions)	(1) Understand laws and regulations. (2) Reflect and implement laws and regulations in actual navigation.
(Influencing factors)	(1) Types of laws and regulations (2) Status of other traffic vessels (3) Conditions in which laws and regulations are applicable (ocean area, weather and sea state, class of ship, principal dimension)
6. Communication (Definition)	Techniques of internal and external communication of intentions by using communication devices such as VHF radio communication systemand intraship telephone.
(Main functions)	(1) Select methods of communication. (2) Prepare information to communicate. (3) Select timing of communication. (4) Use language to communicate.
(Influencing factors)	(1) Communication partners (marine traffic centers, other ships, on the bridge, and onboard) (2) Communication conditions (emergency and standard communication) (3) Devices for communicating (e.g., flashing signal lights, flag signaling, and VHF)

(Continued)

Table I.2.2 (Continued) Definitions, functions, and factors influencing the achievement of functions of the nine elemental techniques

Elemental technique	Details
7. Instrument operation (Definition)	Techniques of effectively using instruments providing relevant information for achieving techniques such as lookout, position fixing, and maneuvering.
(Main functions)	(1) Identify available instruments. (2) Understand methods of using instruments to obtain necessary information. (3) Identify characteristics of information provided by instruments. (4) Identify methods of using provided information.
(Influencing factors)	(1) Types of available instruments (2) Purpose of using information obtained from instruments
8. Emergency treatment (Definition)	Techniques of identifying emergencies with, for example, the main engine or steering gear, or ones arising in the environment surrounding own ship, and performing the necessary actions to respond to malfunctions and emergencies.
(Main functions)	(1) Identify location of problems. (2) Correct problems and malfunctions. (3) Complete necessary tasks related to abnormality and failure occurrences. (4) Identify and take appropriate measures for abnormal behavior in traffic vessels. (5) Identify and respond to abnormal weather and sea states.
(Influencing factors)	(1) Details of emergency (e.g., problems in the hull, cargo, engine, steering system, navigational instruments, and loading and unloading system) (2) Emergencies arising in surrounding ships (3) Abnormal weather and sea states
9. Management (Definition)	Technical management for combining the eight elemental techniques for safe navigation described in the previous sections, and team management for maximizing team member competency to improve team functions.
(Main functions)	Main functions of technical management and team management: (1) Understand management targets (bridge team management, technical management). (2) Understand management methods. (3) Achieve required actions for techniques of management. (4) Evaluate achieved function. (5) Evaluate competency of members related to team management.
(Influencing factors)	(1) Management targets (2) Competency of members related to team management

Figure I.2.2 Organization of the nine elemental navigation techniques

KEY FACTORS OF SECTION 2: DEVELOPMENT OF TECHNICAL ELEMENTS FOR NAVIGATION TECHNIQUES

The necessary techniques for realizing safe navigation are categorized and organized into the following nine elemental techniques:

- Technique of passage planning
- Technique of lookout
- Technique of position fixing
- Technique of maneuvering
- Technique of observing rules of navigation and other laws and regulations
- Technique of communication
- Technique of instrument operation
- Technique of handling emergencies
- Technique of management: managing techniques and team activity

2.1 TECHNIQUE OF PASSAGE PLANNING

Passage planning consists of the technique of collecting necessary information and accordingly creating a safe passage plan. This necessary information can be derived from various sources, including nautical charts, hydrographic publications, port directories, weather forecasts, and tide tables, but the appropriate information corresponding to the navigation area and motion characteristics of the ship being handled (hereafter, "the ship") must be collected. The collected information is recorded as entries in nautical charts and the sailing schedule.

Definition

The planning technique is the technique for collecting information related to the relevant navigational environment and for creating and implementing navigation plans on the basis of this information.

Functions

The passage planning technique is completed when all of the following functions are realized.

(1) **Understanding the information necessary for planning**
All necessary information must be collected. Nautical charts for drafting passage plans, information on traffic separation zones, navigation signals (i.e., information on rules of navigation), and all communications with the vessel traffic information service (VTIS) must be collected. Similarly, other necessary information, such as depth of water related to the draft of the ship, weather and sea forecast, and distance tables, must be collected.

(2) **Understanding the methods of using information**
From the collected information, that which is necessary to meet the navigational objectives of the ship must be selected and processed into effective information that can be used. Inexperienced mariners should ideally not omit any information and must carefully read entries one by one in response to the ship's situation.

(3) **Ensuring that the necessary information is fully reflected in planning**
The information necessary for passage planning is recorded in nautical charts; this is explained in the section titled "Making passage plans". In addition, the appropriate descriptions must be made in other relevant documents, and the critical points should be noted separately for easy recollection.

(4) **Understanding the necessity of modifying plans and making new plans according to the situation**
The original plans may occasionally need to be modified because of various unexpected events during navigation. For example, sudden

changes in weather or sea conditions, abnormal onboard situations, and port closures necessitate new plans that fit the new situation. The drawbacks of the original plans in light of the new situation must be identified, and new plans should be made to address these drawbacks without any delays.

Factors affecting the achievement of functions for passage planning

The objective of passage planning is to ensure that the indicated functions are achieved. The following are examples of the main items that must be considered in planning.

(1) **Traffic rules and regulations**
    Navigation methods and communications with onshore agencies as demanded by the rules and regulations must be considered when making plans.

(2) **Quality and quantity of effective information for navigation (e.g., whether there is a recommended fairway)**
    It is necessary to determine whether the available information is highly reliable. In addition, even if the information is not publicly available, efforts must be made to obtain it from experienced persons. Highly experienced seafarers can anticipate various situations without any information and can collect the relevant information beforehand. Nowadays, there are fewer opportunities for onboard experience, making it harder for seafarers to develop such abilities. Thus, an effort must be made to imagine various situations without relying solely on experience, which might be insufficient.

(3) **Quality and quantity of obtainable basic information (e.g., information on navigation, weather, sea state, geographical features, and waters)**
    There should be a practice of always noting and studying the relationship between obtainable information, navigation plans, and how information is used. The acquisition of information does not occur only at the planning stage. Information is also obtainable at the navigating stage; there are other opportunities to obtain information, and such opportunities must be anticipated.

(4) **Navigation area (ocean navigation, coastal navigation, narrow waterways navigation, fairway navigation, harbor navigation, river navigation)**
    The items that the seafarer must focus on differ depending on the navigating water area. Some examples are consideration of water depth and other marine traffic vessels, the effect of ocean tides on hull movements, the effect of shallow waters, and the bank effect

from the shore. The items affecting maneuvering are explained in the section on ship motion control. Here, in addition to the importance of ship motion control, the importance of verifying pilot support and communicating with information centers at the planning stage is pointed out.

(5) **Purpose of navigation (navigation at sea, dropping anchor, docking)**

In ship handling, the maneuvers when navigating under normal conditions, dropping anchor before or after mooring, or docking and undocking are different, and there are peculiar factors affecting each type of ship maneuvering. The weather and sea state continuously affect the hull movements. In addition, methods for using tugboats and maneuvering under tugboat support should be understood beforehand.

Making passage plans

The important points that must be considered when making passage plans are described.

(1) **Planning for ocean navigation**

Draw planned course lines on large-scale nautical charts and note the approximate travel distance. The following items must be indicated.
- Time under way
- Estimated time of passing at points of altering course
- Estimated time of passing through coastal area and narrow waterways
- Estimated time of arriving at destination

(2) **Planning navigation in coastal areas and narrow waters**

Each of the following items must be noted when passing through coastal areas and narrow waters.

1) No-go areas

Navigable water areas are occasionally restricted by the water depth, their relationship to the ship draft, and the effect of tides. Dangerous areas the ship cannot enter because of such conditions are called *no-go areas*. The establishment of no-go areas is crucial for realizing safe navigation. Thus, the relationship between the ship's position and no-go areas must always be clearly indicated on a nautical chart. Normally, no-go areas are indicated on nautical charts as hatched areas.

2) Safe waters

Considering the following items enables the determination of safe waters, which is one of the bases of realizing safe navigation.
- Draft, maneuverability, and other conditions of the ship

- Precision of navigational instruments and installation of instruments
- Effect of ocean currents, tidal current, and height of tide
- Distance between the bottom of the ship and the bottom of the sea (i.e., under-keel clearance [UKC])
- Distance from shallow water areas and obstacles

3) Under-keel clearance (UKC)

Maintaining adequate UKC is crucial for preventing grounding and ensuring maneuverability, and this must be clearly indicated to ensure safe navigation. From the perspective of maneuverability, if the water depth is less than two times the ship draft, the shallow water effect begins to become influential. That is, in situations in which the distance between the ship bottom and the sea bottom is the same as the draft, there is a decrease in turning performance and changing speed. In particular, in situations in which the UKC is less than 20% of the ship draft, the shallow water effect increases significantly. However, ships with deep drafts, such as VLCCs, are required to navigate water areas with 20% UKC. From the perspective of ensuring maneuverability, a UKC that is at least 20% of the draft is necessary. When the UKC is less than 20%, turning motion, acceleration, and deceleration performance are extremely different from normal; the consequent changes in ship characteristics must be examined beforehand. For ensuring safety, the following points regarding UKC must be considered.

- Accuracy of water depth measurement
- UKC changes with pitching, rolling, and other ship movements
- Effect of squat because of the shallow water effect
- Ship speed, especially rapid acceleration and deceleration

4) Aborts and contingencies

Situations can unexpectedly arise in which a decision on continuing with the scheduled navigation must be made. The boundary of no return must ideally be set in advance in order to make the best decision when encountering such situations. The following points must be considered when determining the boundary of no return.

- The distance to boundary line when entering narrow waters, and the distance to harbor limit lines when entering ports
- Operating condition of components such as the main engine and navigational instruments
- Possibility of receiving pilot and tugboat support
- Weather and sea state

(3) Other general items

General method of making passage plans on nautical charts:

1) Indicate course line and course (course is indicated using three digits).
2) Indicate waypoints for altering course (direction and distance to prominent onshore targets).
3) Indicate steering starting points when there are course changes of large angles.
4) Indicate safe boundary lines, heading target lines, and heading targets in order to ensure safe navigation.
5) Verify the point at which to change to the next nautical chart and the number of the nautical charts to be used next.
6) Indicate the point at which to communicate with VTIS, for example, by using very high frequency (VHF) radio communication systemd the details of communication.
7) Other items to be noted:
   - Areas with congested marine traffic (presence of fishing boats and navigation routes of other ships)
   - Items requiring attention for navigation
   - Sunrise and sunset times
   - Changes in tide

(4) **Items pertaining to entering and departing ports and passage through narrow waters**
1) Stand-by (S/B) engine
   - Schedule the point at which the captain and chief engineer should be called
   - Indicate the starting point of operating main engine
   - Indicate the distance to pilot station
2) Abort points
   - Emergency anchor drop location and water areas for waiting
   - Changes in the route if there is a change in the initial plan
3) Pilot stations
   - Point at which to standby pilot ladder
   - Tugboat mooring method
   - Point at which to assign bow and stern stations

Example of nautical chart descriptions when navigating narrow waters

Figure I.2.3 shows an example of a chart description for navigating coastal zones or narrow waters. The necessary items discussed in the preceding sections are indicated. In this figure, the planned course line is indicated using the thick solid line. At locations slightly off this line, the course is entered using a 360° system in order to keep space for indicating the measured position when deviating from the course line during actual navigation. Altering course points, where course changes are scheduled, are clearly indicated using circles. Next, the distance from the planned course

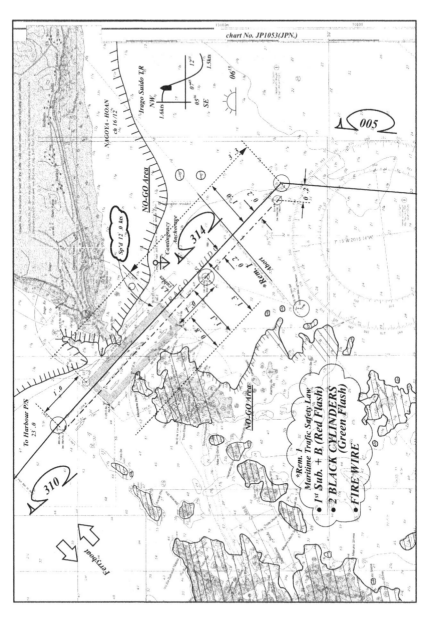

Figure I.2.3 Example of passage planning on a nautical chart

Reproduced from Chart JP1053 with the permission of the Japan coast guard (Permission No. 310001).

line to the targets on both sides and the separation are indicated. This is important for increasing navigation safety and is also necessary for parallel indexing. Next, the no-go areas are indicated, and the limits of the no-go areas set on the basis of the UKC are derived from the relationship of the ship draft and water depth and indicated on the chart. Finally, other points of concern and information relevant to navigating in the planned water area are entered on the nautical chart:

- Expected marine traffic flow to be encountered
- Planned ship speed
- Navigation signal when navigating a specific area
- Tide
- Items related to ship–shore contact

These data are indicated on nautical charts such that they can always be monitored. In recent years, the use of Electronic Chart Display Information Systems (ECDIS) has become commonplace; nevertheless, maintaining the information written on conventional paper nautical charts for constant reference is required for realizing safe navigation.

KEY FACTORS OF SECTION 2.1: TECHNIQUE OF PASSAGE PLANNING

The following functions must be achieved through the technique of passage planning:

1) Collect information for safe passage planning
2) Use collected information to realize safe navigation
3) Perform passage planning in accordance with the collected information
4) Determine when the navigation conditions require changing of plans and accordingly create new plans

2.2 TECHNIQUE OF LOOKOUT

What is generally referred to as *lookout duties* starts with the detection and identification of stationary and moving targets. Furthermore, the present status of the moving targets (i.e., type, distance, direction, and moving speed and direction) must be estimated. This information is essential for predicting future situations, particularly the risks of collision with traffic vessels and their approaches.

Definition

The lookout technique is for detecting and identifying stationary and moving targets, estimating their type, distance, direction, and moving speed and direction, and estimating future risks.

Functions

The lookout technique is completed when each of the following functions is realized.

(1) **Recognition of present situation** (types and movements [i.e., position, course, and speed] of ships encountered)
(2) **Prediction of future situation** (i.e., movement of targets [future position, course, and speed], changes in movement of targets, and estimation of risks to the ship being handled [hereafter, "the ship"]: specifically, closest point of approach [CPA], time to CPA [TCPA], and bow crossing range [BCR; i.e., range from bow or stern])

Factors affecting the achievement of lookout functions

(1) Navigational instruments (compass, RADAR, RADAR/Automatic Radar Plotting Aids [ARPA], Automatic Identification Systems [AIS], and VTIS)

The primary objective of lookout is to detect targets and ascertain their movements in order to avoid colliding with them. Lookout by sight is the basic lookout method, but navigational instruments are currently being increasingly used to improve the lookout capabilities of navigators. However, understanding the unique characteristics of each instrument and their effective use is crucial, as explained in the section on instrument operation. Here, the reason why VTIS is a crucial factor affecting lookout functions is explained. Considering the objective of lookout, obtaining information on the presence and movements of other traffic vessels by using VTIS greatly contributes to the achievement of the necessary lookout functions. Accordingly,

VTIS is expected to yield highly accurate information on the long-term movements of other vessels that is beyond the scope of instrument-based predictions. Therefore, remember that the information provided by VTIS must be used effectively in order to achieve the lookout functions.

(2) **Visibility**

A basic competency required of seafarers is to perform visual lookout and make estimates of the surrounding conditions. However, the visible range is often limited by weather elements, such as fog, rain, and snow. However, even in cases of restricted visibility, ascertaining the surrounding conditions is essential for safe navigation, and the lookout functions must be performed in the same way as when visibility is good (Nishimura and Kobayashi 2009).

Currently, various support systems other than RADAR are available for assisting with lookout when the field of vision is narrow. These support systems have been developed for maintaining lookout functions, because these functions are indispensable for safe navigation.

(3) **Marine traffic vessel volume and traffic flow characteristics**

The most common subjects of lookouts are surrounding ships. When the volume of ship traffic is high in an area, achieving the intended lookout functions may not be possible within the available time. Analyses of seafarer behavioral characteristics have clarified that approximately 10 seconds of steady gazing is required to detect a single ship and to obtain the necessary information. Thus, in the presence of many surrounding ships, repeatedly and continuously monitoring such ships becomes time intensive. Moreover, seafarers may occasionally be in severe situations and must continue travel despite their inability to perform adequate lookout. It is observed that highly experienced seafarers prioritize their lookouts and identify ships presenting a risk of collision at an early stage. They can complete the necessary functions of lookout. Thus, it is important that *ships presenting risks be identified* and *lookouts prioritized*. These tasks are affected by the characteristics of traffic flow in the navigation areas. When the navigation characteristics of the traffic flow are consistent, estimating the future situations becomes easy; this situation requires less lookout time. By contrast, when the traffic flow is not constant, and some ships occasionally navigate differently from the regular flow, dedicated continuous lookout, which is time intensive, will be required.

(4) **Navigation conditions (fairway and traffic laws)**

The general characteristics of traffic flow form spontaneously depending on the arrangement of ports, the shape of water areas, and types of ships. Thus, original traffic flow tends to be inconsistent. In the absence of consistent traffic flow, predicting where ships will approach one another imposes a huge burden on seafarers, and incorrect predictions are occasionally the cause of collision accidents. By contrast, a consistent and regular traffic flow makes it easy to predict the reciprocal behavior of other ships. This is the logic underlying the establishment of fairways and traffic laws. When it is easy to mutually predict ship behaviors, the burden of lookout duties is greatly reduced.

Performing lookout

(1) Identifying present situation

To identify present conditions surrounding the ship, types of surrounding ships and their movements must be known. To this end, their distance and direction from the ship, as well as their course and speed if they are moving objects, must be known.

The method applied for detecting objects and ascertaining their movements is an important factor that warrants discussion. Typically, the achievement level of visual lookout depends on the visibility conditions, but it is considered effective; *visual lookout may be more easily achievable with fewer errors than in other methods.* Furthermore, certain target types and movements can be detected through careful visual observation. With the recent development of RADAR/ARPA and ECDIS, the function of lookout tends to use these instruments. The advantage of these instruments is that even seafarers with little lookout competency can determine target movement through a review of the numerical data. However, remember that instrument-based lookout provides limited information compared with visual lookout. In addition, depending on the characteristics of the instrument, there may be occasional delays in sensing, and some movements may not be detected.

In cases of extremely restricted visibility, instrument-based lookout becomes inevitable. However, always remember that even in such cases, the information acquired may be limited, as explained earlier. Furthermore, studies have demonstrated that the lookout range of RADAR/ARPA under limited visibility is smaller to that under good visibility. The important functions of lookout are identifying situations, predicting future situations, and analyzing information in order to determine the necessary actions, all of which require time for information processing. Therefore, *early first detection is essential to ensure adequate time for executing the determined actions.*

However, when the lookout range is limited, objects may be detected late, making necessary predictions impossible or resulting in delayed avoiding action, which may lead to near misses or even collision accidents. The lookout range changes because of the various aforementioned external conditions. Thus, the lookout range directly influences the time to detect dangerous targets.

One factor affecting the lookout range is the number of surrounding ships. When many vessels surround the ship, seafarers must constantly observe the movements of these vessels. This results in the lookout range being concentrated on surrounding ships and not on a wide area. Figure I.2.4 illustrates the relationship between the number of vessels surrounding the ship and *the time of first detection of a target, which is defined as the time to the CPA of a target* (TCPA). The horizontal axis represents the number of surrounding ships, and the area defined as the surroundings, a 3 mile radius circle centered on a point 2 miles ahead of the ship, is indicated in Figure I.2.5. The center of this circle is 2 miles ahead of the ship because regular seafarers look out for other ships primarily in the heading direction. Four ships can be seen in the circle in the example in Figure I.2.5. As shown in Figure I.2.4, in the absence of any surrounding ships, an approaching vessel can be first detected at a TCPA of approximately 30 minutes, meaning that a ship presenting a danger can be noticed 30 minutes before the CPA. By contrast, when seven ships are present in the vicinity of the ship, the first detection is delayed until 15 minutes before the CPA. If there are 30 minutes until the CPA, the available time is adequate to continuously monitor changes in the situation after first detection, to appropriately estimate future situations (as explained in the next section), and to accordingly propose and execute the optimum collision-avoiding action. However, the time available for these actions will be inadequate if the TCPA at first detection is 15 minutes or less, and determining the optimum collision-avoiding action becomes difficult (Nishimura and Kobayashi 2005).

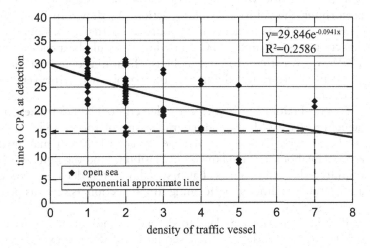

$$y=29.846e^{-0.0941x}$$
$$R^2=0.2586$$

Figure I.2.4 Relationship of surrounding ship density and TCPA at time of first detection of target

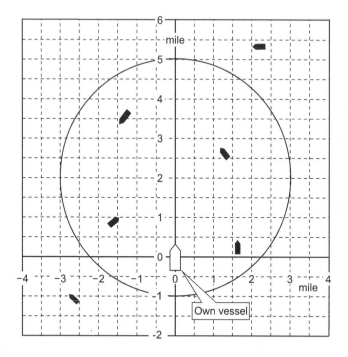

Figure I.2.5 Definition of surrounding ship density

Figure I.2.6 illustrates the relationship between the time of first detection and the results of maneuvering to avoid collisions with a target. The horizontal axis indicates the TCPA when a target is first detected, and the vertical axis indicates the distance at CPA after performing the collision-avoiding action. As shown in Figure I.2.6, a target would approach within four cables that is 0.4 mile if first detection was 10 minutes TCPA, and there would be a collision if the TCPA was less than 5 minutes. Thus, the initial detection time is a factor in determining whether collision-avoidance maneuvers would succeed.

Several factors can hinder early detection. The tendencies in first detection discussed here are characteristic of the behavior of standard seafarers. In other words, even normal behavior may produce risks. This book has explained how visibility and the number of surrounding ships are factors that may delay initial detection. As these are routine conditions, performing first detection as early as possible becomes crucial.

Routine behavior may lead to changes in lookout range, so a conscious effort at early detection should be made. For example, there should be a practice of periodically expanding the range of RADAR and of searching for distant targets using binoculars.

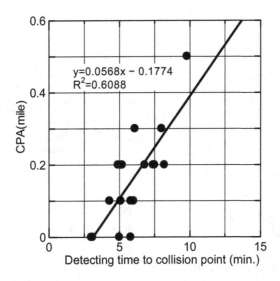

Figure I.2.6 Relationship between first detection time of risky ships and CPA

(2) Prediction of future situations

The primary purpose of the lookout technique is the collection of information to avoid collisions with objects. The necessary information must be acquired to perform appropriate collision avoidance. Presently, systems such as AIS have been developed and are being used to create conditions in which it is easy to make estimates and predictions regarding the behavior of other ships at sea. However, because *updates to AIS information* are occasionally not renewed and because *detailed movements are not informed by AIS, making predictions through lookout remains a basic necessary technique for safe navigation.*

Information on the movement of objects obtained through lookout includes present position, course, and speed. On the basis of this information, predictions of future risks to the ship are made; subsequently, appropriate countermeasures are determined if necessary. Information crucial for these predictions are CPA, TCPA, and the distance to another ship when it passes the bow or stern of the ship (i.e., bow crossing range (BCR)). CPA and BCR indicate the separation when two ships are approaching each other.

BCR is important for planning maneuvers. Figure I.2.7 shows that when BCR is positive, the course of the ship must be changed in the direction of the other ship in order to widen the separation. However, when the BCR is negative, the other ship would pass the stern of the ship, so a course changing toward the other ship would result in a

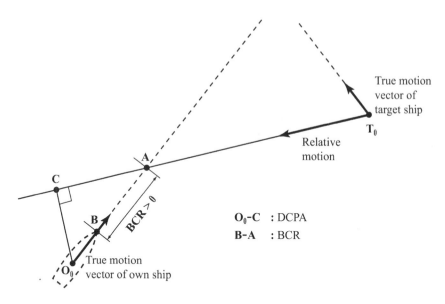

Figure I.2.7 Relationship between BCR and DCPA

closer approach. Thus, *BCR directly indicates the actions* that must be taken by the ship.

Seafarers with little experience often rely only on CPA to ascertain future situations. *CPA indicates the degree of safety as the result of avoiding action.* Therefore, CPA alone should not ideally be used to determine maneuvering plans. Predicting and determining whether the bow or stern would be passed should be primarily performed through visual observation, the recommended technique.

When the visibility does not allow visual observation, the information provided by RADAR/ARPA, in which future risks are displayed as numerical values, becomes valuable. Observations of the behavior of seafarers, inexperienced seafarers in particular, have revealed their high tendency to depend on instruments to obtain crucial information. However, the dangers of overreliance on instruments should be remembered. The information displayed by navigational instruments might be slightly delayed. Furthermore, predictions provided by the instruments are estimates based on this delayed information. Consequently, the presented information and predictions might not be true of actual situations, and the occasional incorrect prediction must be recognized.

Next, the standard process for avoiding collision risks is described.

(3) **Process in lookout from first detection to action**

The key requirements for lookout are to understand at what time other ships should be detected and by what time a risk of collision should be identified. In other words, performing safe avoidance

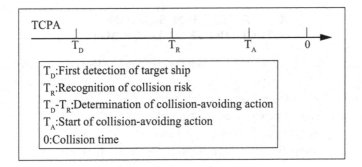

Figure I.2.8 Progress of collision-avoidance action

maneuvers becomes impossible if the seafarer cannot detect the target or recognize the risk before a certain TCPA. This information must be known before starting any collision-avoiding action. Figure I.2.8 is a horizontal line showing the elapsed time corresponding to TCPA at each process. Here, T_D, the time at detection, is the TCPA when target ships are first detected (i.e., the time until collision after a target is first detected); T_R, the time at recognition, is the TCPA when a collision risk is recognized; and T_A, the time of starting action, is the TCPA when a collision-avoiding action is initiated. The point where TCPA is 0 indicates the time when a collision will occur if no collision-avoiding action is taken. T_A should therefore correspond to the time required for completing the collision-avoiding action, during which a safe separation distance must be ensured. The necessary time for completing the collision-avoiding action differs depending on the encounter conditions and the maneuverability of the ship. The time from T_D to T_A is the time required to collect and process the lookout information, and this time is independent of the encounter conditions and maneuverability.

According to an international joint study by the International Marine Simulator Forum conducted between 2003 and 2006, the time from T_D to T_A is approximately 15 minutes. Consequently, the minimum TCPA at first detection can be determined by adding 15 minutes to the time T_A required for collision-avoiding action:

$$(\text{TCPA at first detection})$$

$$= (\text{time required for collision} - \text{avoiding action}) \qquad (2.1)$$

$$+ 15 \text{ minutes}$$

The necessary T_A is dependent on various conditions and has been reported to be approximately 10 minutes on average. Consequently, the time of first detection of a target should be 25 minutes before CPA.

KEY FACTORS OF SECTION 2.2:
TECHNIQUE OF LOOKOUT

Functions that must be achieved using the technique of lookout are as follows:

1) Understanding of current conditions
   - Early detection of other ships
   - Types of target ships encountered
   - Movements of target ships encountered (position, course, and speed)
2) Understanding of conditions that the ship being handled will encounter in the future
   - Future conditions of target ships encountered (future position, future course, and future speed)
   - Risks of collision (CPA, TCPA, and BCR)

2.3 TECHNIQUE OF POSITION FIXING

This technique involves functions to estimate the present position of the ship being handled (hereafter, "the ship") by using visual observation or navigational instruments. Tide and wind influence future positions of the ship and thus are factors affecting movements of the ship that must be ascertained as functions of position fixing.

Definition

The technique for estimating positions of the ship through optimal object selection and visual and instrumental methods as well as techniques for estimating the factors influencing movements of the ship.

Functions

(1) **Selection of methods to collect information for position fixing**

Information can be collected through both vision-based methods and navigational instruments such as Global Positioning System (GPS) and RADAR. The selected methods should be highly reliable. Furthermore, position fixing should be performed using as many different methods as possible. In addition, whatever the methods, the measurement objects and information must be carefully selected. The development of navigational instruments in recent years has led to their widespread use for collecting information. In particular, ECDIS, which uses GPS information, is being increasingly used to verify ship position. Consequently, competency in position fixing through visual observation and by using RADAR information has decreased. However, an unexpected increase in sunspot activity in recent years has led to the issuance of warnings of electromagnetic interference. This means that there may be unpredictable situations in which ship position cannot be verified using ECDIS because of the unavailability of GPS signals. Moreover, such situations may arise during navigation in narrow waterways. Therefore, high competency in vision-based position-fixing techniques, which have been a basis of navigation, should be maintained.

(2) **Ship position determination**

When determining the ship position, the required precision of the estimates and the frequency of position fixing differ depending on the navigation area and navigational conditions. Compared with travel in open waters, navigating narrow waterways and fairways requires more precise and more frequent measurements of ship position. In particular, ship position must be frequently verified in water areas where the effects of wind pressure and tide on ship movements are high.

Even in conditions requiring frequent measurements, there is an upper bound on the practicable frequency. Thus, efforts must be made to reduce the time and steps necessary for each measurement. The methods for setting heading objects, ascertaining the positional relationship to important objects, and using the obtained information must all be considered at the passage planning stage itself. Parallel indexing is an important technique for continuously ascertaining ship position.

(3) **Estimation of movements of the ship being handled**

One of the objectives of position fixing is the estimation of the actual movements of the ship. This function, competency in which is often lacking in inexperienced seafarers, is crucial when navigating under the effect of wind pressure and tides. Setting courses by ascertaining and accounting for the influence of external forces is a vital function for safe navigation.

In addition, estimated time to arrival (ETA) is fixed after determining the time when the ship enters the fairway and approaches a pilot station. Occasionally, ship velocity must be adjusted moment by moment in order for the ship to arrive at its destination at the fixed time. In such conditions, accurate position fixing and speed adjustments based on the position are important; this function is performed as a movement estimation function of the technique of position fixing.

Factors affecting the achievement of position-fixing functions

The feasible position-fixing precision and frequency are dependent on the following conditions, which affect the achievement level of position-fixing functions.

(1) **Types of instruments available for position fixing**

In addition to visual position fixing, in recent years, systems such as ECDIS have been installed on many ships, enabling easy position fixing. ECDIS displays moment-to-moment ship position on electronic nautical charts and is thus an extremely convenient instrument for understanding the current relationship between the geographic condition and the ship position. Position measurement is based on the analysis of GPS-derived information. Occasionally, position fixing is performed through RADAR imaging–based position measurement or echo sounder–based depth measurement when the ship is close to the shore. The optimal method for the circumstance should be used, and the required position-fixing precision and frequency must be considered.

(2) **Navigational environment conditions**

Position fixing is becoming increasingly reliant on navigational instruments. However, when performing the technique for the lookout of surroundings, visually estimating the ship position by using visually obtainable fixed objects (e.g., lighthouses, cape geographic features, onshore structures, and island silhouettes) is crucial for safe navigation. During narrow waterway navigation, lateral deviation from the planned course can be easily detected by setting the heading objects. Thus, during passage planning, considerations that facilitate position fixing are important.

(3) **Visibility**

Visually making moment-to-moment estimates of the ship position is a basis of safe navigation. However, the conditions in which vision-based position fixing is to be performed can change: visibility can be limited by fog, rain, and snow. In such cases, a ship position measurement method that is highly reliable under the given visibility conditions must be used.

(4) **Disturbing elements**

The ship must be constantly monitored for unintended hull movements and deviations from the planned navigation route because of the effect of tides and winds. Hull movements due to wind pressure increase gradually, but movements due to tide can affect the ship immediately and thus must be managed immediately. During fairway navigation, the position-fixing frequency is dependent on the width of the fairway and the tidal speed and direction. Position fixing must account for these elements, and the heading must be corrected accordingly.

KEY FACTORS OF SECTION 2.3: TECHNIQUE OF POSITION FIXING

Functions that must be achieved by the technique of position fixing are as follows:

1) Select the method of information collection for position fixing.
2) Estimate the ship position on the basis of the collected information.
3) Evaluate the ship movement and estimate the information necessary to evaluate the effect of external disturbances and to realize passage planning.

2.4 TECHNIQUE OF MANEUVERING

This technique is for realizing planned movements of the ship being handled (hereafter, "the ship") and thus achieving safe navigation on the basis of adequate understanding of maneuvering characteristics. Furthermore, techniques for operating the main engines, rudders, side thrusters, and tugboats on the basis of knowledge of the effects of water depth, wind pressure, and tide on hull movements are covered.

Definition

The technique of maneuvering involves functions for controlling the course, speed, and position of the ship by, for example, operating the rudder and controlling the main engine.

Functions

The following functions must be achieved when performing the maneuvering technique.

(1) **To measure the ship motion**
     First, the movements of the ship must be ascertained in order to control it as planned. The motion-related factors to be measured depend on the intended movement:
   - Course control: Yawing angle (turning angle) and rate of turn
   - Controlling the ship position: Yawing angle (turning angle), rate of turn, ship position (latitude and longitude), and ship moving speed
     Proportional integral differential (PID) controllers are employed for basic motion control. In PID, three elements (proportion, integration, and differential yield) control variables that can be used to measure and accordingly control ship movement. For the proportional control element (P), the difference between the intended and actual movement is a control variable. For example, the aforementioned difference in the rudder angle, shown in Equation 2.2, is constantly corrected to maintain the planned course.

Rudder angle = Proportional coefficient

$$\times (\text{planed course} - \text{present course}) \qquad (2.2)$$

If the rudder angle is determined to be proportional to the difference between the target and the actual course, then the rudder angle is steered such that the ship moves toward the opposite of the actual course. Consider that the target course is 030° and the current course is 031°; by using Equation 2.2 and assuming that the proportional coefficient is 1.0, the rudder angle is changed by $-1.0°$. Thus, through proportional control, if there are differences between the target and the current heading angle, a counteracting operation will be applied until the difference reaches 0.0. However, when the difference becomes 0.0, the turning motion will not stop even if the current rudder angle is 0.0, and the heading angle will overshoot the target course, creating a heading error on the opposite side of the ship. Then, a counteracting operation will be applied in the opposite direction to reduce the overshoot angle error. Thus, the course continuously fluctuates around the target value. A large proportional coefficient decreases the time required to reach the target value but increases the heading overshoot error range, whereas a small proportional coefficient reduces the heading overshoot error range but increases the time required to reach the target value.

With proportional control, movements can occasionally overshoot the target values, and the corrective actions can be slow, requiring a long time to reach the target values. To overcome these disadvantages, the differential (D) control method is employed to account for changes in speed from the present to the target movement. If the speed of change is too high relative to the error difference, then the operational load must be reduced to fall within a range in which the target values are not exceeded. To enable D control in which the speed of, for example, yaw rate change is reflected as a control variable, the moment-to-moment speed of yaw rate movement is measured and serves as the control variable. Subsequently, the yaw rate corresponding to the speed of heading change in the turning angle becomes the amount of movement measured.

The control value is applied under the assumption that the resulting movement arises solely as a result of one's own operations. However, in actual environments, movements can simultaneously be induced by wind and tide. When movements deviate from the target values because of external influences, the corrections made through proportional and derivative control are late, which results in unnecessary movements. In such cases, integral (I) control, which compensates for movements arising from external forces in advance, is effective.

During actual ship maneuvering, the heading angle can be easily determined from compass readings. Then, how should the rate of turn be measured? The accelerating and decelerating turning movements of large vessels such as VLCCs are quite slow. Hence, changes in the control value cannot be applied at an accurate time solely through compass observations, making correct derivative control impossible. Thus, rate-of-turn indicators must be installed on large vessels to supplement accurate human-based movement measurements.

(2) **Selection of control device**

The rudder and main engine can each be operated separately, and simple course and speed adjustments can be made without much difficulty. However, as ship size increases, ship motion measurement and maneuvering become difficult.

Several systems, such as side thrusters installed on the ship, tugboats supporting ship motion control, anchors used for mooring force, and springing lines for adjusting the mooring forces as well as the main engine and rudder are applied to maneuver the ship by controlling its motion.

Next, several maneuvering situations are considered. First, let us discuss docking using a tugboat. In this case, in addition to the ship's rudder and main engine, multiple tugboats must be in operation. Five tugboats assist each VLCC that docks at Keiyo Sea Berth in Tokyo Bay. The ship movements are controlled by issuing commands to each tugboat regarding its force, direction, and position relative to the ship. Three elements must be determined in order to control each tugboat: the position of, force generated by, and direction of force of each tugboat. Thus, when using five tugboats, 15 elements must be constantly ascertained. In addition, six motion elements pertaining to the current movement relative to the desired movement must be understood: heading angle, longitudinal position, lateral position, rate of turn, longitudinal speed, and lateral speed. Thus, in total, 21 parameters must be constantly ascertained to determine the control variables.

Human memory and information-processing capabilities are limited, so the actual controls are adjusted such that the number of control elements can be reduced to a processable range. Effective tugboat use and safe docking are closely related to the characteristics of human control, an aspect that will be discussed later in this book.

The methods for controlling the ship are chosen depending on the situation. When controlling heading, the steering rudder is applied: one device is used for controlling one motion. However, complicated motions entail multiple devices.

Consider the case of approaching a pilot station while adjusting the estimated ETA. In this situation, the ship position must be adjusted on approach. The operating input for this control is the rudder angle, and the movement to be controlled is the heading and position. Furthermore, main engine operation is involved in adjusting the speed in order to maintain the ETA. The number of movements to control, as well as the operating inputs, varies depending on the maneuvering situation. A complication of ship maneuvering is that the main engine and rudder operations are not mutually independent; the operational situation of the main engine can effect changes in the rudder force. Situations commonly arise, especially during ship maneuvering, in which the rudder and main engine must be operated simultaneously for course and speed control, respectively. When such dual systems are in simultaneous operation, the movements produced by the two systems may mutually interfere, complicating the maneuvering and occasionally increasing the operational burden on the operator. Maneuvering difficulties particularly increase when the effects of wind, waves, and tides are high. Appropriate maneuvering requires determination of which ship equipment to operate at each moment and subsequently its appropriate operation. These complicated operations are bases of safe navigation and highlight the importance and difficulty of the techniques collectively referred to as maneuvering.

(3) **Determination of value of operation**

As described earlier, even in cases of course control, which is the simplest maneuvering situation, acceleration and deceleration movement is quite slow for VLCCs. If the operation value is not determined at some moments during the slow movement, the target course may be overshot, or the movement may be stopped, and steering repeated. Thus, acceleration movements should be determined on the basis of the ship size and the operation value. The operating force achieved through rudder operation does have limits, so for VLCCs and other large vessels, only slow movements can be realized.

On the other hand, if the control target is small and the control force is large, another problem arises: a large rudder force is applied to the ship hull with little operation, resulting in rapid movement. When force is applied to arrest this movement, the movement stops immediately but is then generated in the opposite direction. In general, systems in which the control force is large are more difficult to operate than are systems of large vessels; such systems are thus called *unstable systems*.

Course control has been described as a typical example of a single input, single output operation; however, ship maneuvering is not simple even for such operations. As described earlier, the number of operating inputs and the volume of movements necessary for ship control vary with the situation. In sum, in ship maneuvering, the selection of ship operation equipment and the determination of operation value are critical technical actions.

Factors affecting the achievement of maneuvering functions

(1) **Maneuvering objectives**

Ship maneuvering must be performed in many situations, ranging from simple course control to docking with the support of multiple tugboats under the effect of tidal currents. Maneuvering difficulty changes with the situation, as does the degree of intended maneuvering achieved.

(2) **Available control devices**

Single input, single output control systems allow easy control. However, the primary ship maneuvering is performed at ports and involves multiple inputs and outputs, making for complicated and difficult maneuvering. Maneuvering does not get easier with an increase in the number of available control devices. The ideal setup is one in which systems that people can use to easily ascertain situations are realized. That is, if the same movements are being controlled by a single input device, the necessary operations are simple; thus, system designs that simplify operational load determination are desired.

(3) **External forces**

Ship navigation, particularly maneuvering techniques, is more directly and strongly affected by environmental influences than are other traffic systems. The purpose of maneuvering, as defined for seafarers, is to realize the planned ship movements by using the control systems. Control systems apply force to the ship's hull in order to produce the intended movement; however, the external forces that influence hull movement are uncontrollable. Occasionally, these external forces can be utilized, but their effects may also be opposite to those desired. In addition, it may occasionally be impossible to produce the necessary movements because of the external forces. For example, course cannot be maintained under the effect of strong winds if the turning moment generated by the wind pressure exceeds the maximum turning moment that can be implemented using the rudder; in this case, the intended course cannot be maintained even if the largest rudder angle is applied.

Wind pressure and tide are the other major factors that affect hull movement during harbor maneuvers, narrow waterway navigation, and docking; these factors significantly influence the maneuvering functions (Ishibashi and Kobayashi 2009.)

KEY FACTORS OF SECTION 2.4: TECHNIQUE OF MANEUVERING

Functions that must be achieved by the technique of maneuvering are as follows:

1) Measure and ascertain current movements of the ship.
2) Select instruments to operate in order to realize the planned ship motion.
3) Determine control variables of control systems to realize the planned motion.

2.5 TECHNIQUE OF OBSERVING RULES OF NAVIGATION AND OTHER LAWS AND REGULATIONS

This technique involves the knowledge of traffic laws related to navigation, including the 1972 International Regulations for Preventing Collisions at Sea (hereafter, COLREG), the Local Traffic Safety Act, and the Act on Port Regulations, and the ability to apply it to actual navigation. This technique also includes the knowledge and application of local rules in each area of navigation.

Definition

Technique for navigation according to rules such as COLREG, the Local Traffic Safety Act, and the Act on Port Regulations

Functions

(1) **Understanding laws and regulations**

Detailed explanations of the various rules of navigation can be found in the many published textbooks and thus are omitted from this book. Nevertheless, seafarers must have a thorough understanding of the conditions under which the laws and regulations are applicable.

(2) **Techniques for actual navigation based on application and implementation of laws and regulations**

Merely understanding and remembering the relevant laws and regulations does not mean that they will be appropriately applied in actual navigational situations. Putting knowledge into actual practice is an important requirement for all techniques.

Novice seafarers often exhibit certain behaviors during maneuvering training. Commercial vessels are usually large in size, and collision-avoiding actions are initiated 5 miles or farther from other ships, primarily to ensure safety. The necessary actions are better taken while there is some distance between the ships than when closing in on approach. If action is initiated when the ships are largely separated, COLREG does not necessarily apply. However, novices tend to be fixated on following the regulations, and the resulting actions might not actually be the safest or most appropriate option. On the other hand, in several actual cases, even though the COLREG regulations were understood, dangerous collision-avoidance maneuvers in violation of the laws and regulations were initiated within 3 miles, where there is coverage by COLREG. In other cases, novices often steered the heading to that of the other vessel at close range after overtaking the vessel.

Even experienced seafarers are occasionally found lacking in knowledge of the rule in water areas where they do not have much regular navigation experience. Recent years have seen a rise in the numbers of seafarers who have spent few years at sea and those with much experience in navigating specific types of ships on specific routes, leading to a general lack of sufficient knowledge of traffic regulations along other routes.

Factors affecting the achievement of law and regulatory observance functions

(1) Types of laws and regulations

Seafarers must be well aware of the laws and regulations applicable to the water areas being navigated. In cases of conflicting rules, the appropriate order of priority of application of laws and regulations must be followed.

(2) Status of other traffic vessels

Traffic regulations are implemented to ensure the safety of marine traffic vessels. The behavior of the ship being handled is always determined by that of other ships. Adequately monitoring the status of other ships and constant awareness and application of the relevant laws and regulations are both crucial. Some ships may not follow the rules, so attention and warning signals should be effectively used. In addition, in congested water areas with ship traffic or fishing vessels, navigating according to the laws and regulations may occasionally be difficult and may be limited by ship maneuverability and draft, thus requiring the cooperation of other traffic vessels to ensure safe navigation. In such cases, whenever time permits, such cooperation must be sought through VHF wireless telephone or other means, or attention signals must be used.

(3) Conditions under which laws and regulations are applied

The conditions under which laws, regulations, and rules are applied differ depending on the conditions of the navigational area, weather, and sea. These conditions must be understood, and navigation performed accordingly.

KEY FACTORS OF SECTION 2.5: TECHNIQUE OF OBSERVING RULES OF NAVIGATION AND OTHER LAWS AND REGULATIONS

Functions that must be achieved using the technique for rules of navigation and other laws and regulations are as follows:

1) Understanding the relevant laws and regulations
2) Reflection and implementation of the laws and regulations in actual navigation

2.6 TECHNIQUE OF COMMUNICATION

This technique corresponds to the technique for onboard and offboard exchanging of mutual intentions through communication means such as VHF radio communication system and intraship telephone. The scope of this technique is wide and includes the use of whistles, signal lights, and searchlights.

Definition

The technique of exchanging information with internal and external places through VHF radio communication system and intraship telephone

Functions

(1) **Selection of methods of communication**

Communication between crewmembers on the bridge is essential when multiple seafarers collaboratively perform the tasks necessary for ship navigation. This function is a crucial part of bridge team management and is thus detailed in that chapter. In such situations, communication is mainly oral.

Communication outside the ship being handled (hereafter, "the ship") is mainly through VHF radio communication system; nevertheless, the importance of effective use of whistles and signal lightsmust be understood, particularly when communicating the ship's intention to fishing boats or small ships, which may not be equipped with VHF radio communication system. If a seafarer feels the need to call for attention or is unsure of the behavior of other ships, whistles and other signals must be effectively used.

(2) **Content of communication**

The purpose of communication is to exchange each intention. The intention of the transmitter must therefore be conveyed clearly to the receiver. If the receiver has correctly understood the intentions, it can then reply appropriately. To achieve the objectives of communication, these objectives must first be clarified by using simple but complete sentences that exchange sufficient information.

Moreover, stating the reasons for sending information will ensure clear understanding of the transmitted information. For example, when confirming the other ship's navigation plans, the following communication steps are recommended:

- Send the ship's information necessary for the other ship (i.e., the receiver) to recognize the transmitter.
- State the ship's navigation plan.
- State the objective of communication of intention and confirm the other ship's navigation plan.

If the interaction is confirmed after exchanging this communication, the communication is continued as follows:

Example 1: Ask the target vessel to take collision-avoiding action because inadequate space is available for the ship to take collision-avoiding action as a result of the water area restrictions.

Example 2: Confirm the collision-avoiding actions agreed upon by both vessels.

The objective of communication is the mutual understanding of intentions between ships. Therefore, the communication must be clear: the transmitter must always ensure that the communication will be easily and accurately understood by the receiver and must clearly state the purpose of communication.

(3) **Selection of communication timing**

The timing of communication with other ships through VHF radio communication system is important, particularly when communicating to avoid collisions. In a collision-avoidance situation, the main purpose of communication is usually to verify mutual intentions in order to determine the appropriate collision-avoiding actions. To realize safe collision avoidance, after the mutual intentions have been verified, *adequate time must be available for completing the collision-avoiding action*. Furthermore, communication may occasionally conclude unsuccessfully, or collision avoidance must be taken on the basis of the ship behavior. In light of these discussions, *the time when a collision risk has been recognized* is the ideal time for establishing communication.

(4) **Language for communication**

Onboard communication during international voyages is primarily in English. The IMO has published books on standard marine communication phrases and other related subjects.

In the author's experience of educating and training, both navigators and students with proficiency in English do not always attain acceptable navigation communication.

Most importantly, the *objective of communication* must be effectively achieved. Thus, the conversation must match the needs of the situation. In other words, seafarers must *appropriately judge the situation and accordingly communicate clearly and concisely*. What is being communicated differs in standard and emergency situations. If a navigator with little experience or low English conversation abilities can appropriately judge the situation, then they can achieve the communication objectives by employing basic communication methods.

Factors affecting the achievement of communication technique

(1) Communication partners

The targets of communication include crewmembers assigned to the bridge, other crewmembers aboard the ship, crewmembers of other ships, and staff at marine traffic centers. *Communication with crew assigned to the bridge* is a crucial function, and the methods of communication, the information communicated, and the communication timing are addressed in the section on bridge team management in Part II of this book. Communication may be with crew assigned to the bow and stern, and to the engine room, as detailed in Part II. In this part, I have highlighted the necessity of effective communication from the perspective of ship team management.

A typical example of *offboard communication* is the reporting of passing predetermined locations and the ETA at a specific fairway to marine traffic information centers. The main information to be communicated is determined by the relevant center. It is to the benefit of the ship to proactively obtain crucial navigation information from the traffic centers; that is, centers can be used as information sources.

(2) **Communication situations**

Communication situations can be either emergencies or standard situations. To ensure preparedness *for emergency communication*, the communication content and communication partners must be determined and prepared beforehand. Emergency situations are a race against time, and efficient error-free methods of communication must therefore be prepared during standard situations. *Even under standard operation situations*, inexperienced seafarers must make preparation for the possible situations beforehand.

(3) **Methods of communication**

When VHF radio communication system is used for communication outside the ship, the ship's intentions must be accurately signaled to the communication partners, particularly when the communication partner is a ship. In recent years, the use of AIS has eased the identification of communication partners. When AIS-based identification of other ships is difficult, calling methods must be performed for accurate identification. For example, in addition to the other ship's position, course, and speed, ship type, hull color, and other characteristic structures can be used as calling elements. In addition, when the purpose of communication is to clarify mutual behavior, the behavior of the other ship must be observed carefully even after the communication is complete in order to verify whether the actual movements match those communicated. This verification step is essential because occasionally, the actual communication partners can be different from those intended or the communications may not have been clearly understood. The signal flag, air phone, and searchlight are useful devices to show the intention to vessels in the vicinity. The communication methods must be selected according to the target conditions.

KEY FACTORS OF SECTION 2.6:
TECHNIQUE OF COMMUNICATION

Functions that must be achieved by the technique of communication are as follows:

1) Select methods for communication.
2) Understand how to engage in communication.
3) Understand when to engage in communication.
4) Understand how to apply proper language.

2.7 TECHNIQUE OF INSTRUMENT OPERATION

The purpose of the technique of instrument operation is to effectively select and use the navigational instruments installed on the ship being handled (hereafter, "the ship"), particularly on the bridge, in order to acquire the information necessary for safe navigation; that is, this technique covers the use of instruments and the information obtained using the instruments.

Definition

Techniques for effectively utilizing instruments providing relevant information for achieving lookout, position fixing, maneuvering, and other techniques.

Functions

(1) **Recognition of available instruments**

Many maneuvering instruments and instruments providing information beneficial to navigation are installed on the bridge. Instruments for controlling ship movements include the steering wheel, the main engine control system, and the side thruster control system. Other instruments, excluding these, provide information. Over the long history of shipping, techniques available during certain eras have been used to develop and use various instruments. With scientific and technical progress, the quality and quantity of the information obtained from the instruments have improved. Ships contain both the latest and most advanced instruments as well as older instruments, meaning that the same type of information can be obtained using multiple instruments. In addition, in some instruments, the same information can be obtained through different measuring methods (e.g., direction information from magnetic compasses and gyrocompasses). Seafarers must use the most reliable instruments for obtaining the necessary information. In addition, when an instrument appears to be malfunctioning, a suitable alternative instrument must be identified. Thus, seafarers must constantly be aware of what information is being obtained through which instruments. Moreover, the precision and reliability of the information provided by the different instruments may not always be the same, so seafarers must understand the characteristics of the information they are using.

(2) **Understanding methods of using instruments to obtain the necessary information**

The methods of using each instrument are different; therefore, seafarers must understand how to obtain the necessary information: the various functions of each instrument and the method of using each function must be understood. Therefore, the technique of utilizing the instruments and the information is crucial for safe navigation by application to the relevant techniques, which are lookout, position fixing, maneuvering, and so on. Here, the various instruments are categorized according to the information they yield, and the way to use these instruments will be described.

1) Acquiring information on traffic vessels

Methods for obtaining information on surrounding ships by using instruments include RADAR/ARPA, AIS, and communication. Methods for using information from communication were explained in paragraph 3), so the other instruments are discussed in this section.

The functions to be used when information on other traffic vessels is to be obtained using RADAR/ARPA include *selection of the optimum measurement range, selection of the motion vector length, and determining whether the display shows relative or true movement.* The measurement range corresponds to the lookout range: A short range is sufficient when monitoring surrounding ships, but faraway objects require periodic long-range monitoring. Delayed target detection can result in delayed initiation of collision-avoiding actions, which increases the risk of collisions and near misses. If there are few surrounding ships, then the first detection must be at least 25 minutes before CPA. Estimation of the approach conditions and collision-avoidance methods through continuous monitoring after first detection is a basis of safe navigation. Subsequently, collision-avoiding action must be initiated with at least a 10 minute lead before the CPA.

Even in situations in which the surrounding ship density is high, efforts at early detection must be made to ensure the aforementioned 15 minute lead. Thus, the range used for RADAR/ARPA monitoring of surrounding ships must be frequently varied in order to detect long-distance targets early.

When using RADAR/ARPA, how each function is utilized is important. For example, consider the use of motion vectors for determining the speed and heading direction of a moving target. If the length of a motion vector is short, not only is it difficult to make predictions, but it is also hard to detect changes in movement. The recommended length of motion vector is the same as the range used: for example, if the range used is 6 miles, the length of the motion vector should be 6 minutes. The use of vectors that are too long relative to the range is rare, but situations often arise in which short motion vectors are used even when the measurement range is long. Compared with relatively long vectors, the use of relatively short vectors more strongly hinders the monitoring function.

CPA or distance at closest point of approach (DCPA) is typically used as a measure of the collision risk between marine traffic vessels. However, accurate judgments cannot be made solely using CPA or DCPA, because they do not include elements concerning the positional relationship of both ships when they are at the CPA. DCPA can be used to accurately determine the effects of collision-avoiding action as a result, but the attitude of each vessel when passing is important for actual maneuvering, so CPA alone is insufficient. Therefore, the latest RADAR/ARPA systems display the separation when the target ship passes the bow or stern, that is, the BCR. A positive BCR indicates that the target ship would pass the bow of the ship, whereas a negative BCR indicates that the target ship would pass the stern. BCR thus critically influences the decision making pertaining to collision-avoiding actions. However, the BCR is not displayed for all ships on the RADAR/ARPA system; hence, the BCR must be obtained through other methods. One method is to display vectors in the relative mode, which extends the speed vectors of the other ships; these vectors can be used to determine whether the target ship would pass the bow or stern of the ship. This information can also be derived from true vectors by extending the true vector of the ship and of the target.

RADAR/ARPA has other functions as well, such as fixed range marker display, offset center display, and collision-avoiding action simulation; therefore, a thorough understanding of these functions and their use is essential for safe navigation.

The aforementioned functions are closely related to the lookout functions; specifically, effectively using the instruments can improve the lookout functions. Instruments have been developed and improved for supplementing the essential navigation functions performed by seafarers.

2) Verifying the ship position for safe navigation

In addition to collision accidents, grounding accidents must also be avoided to realize safe navigation. To this end, monitoring geographic features and their relative position from the ship is vital. The recent development of ECDIS using GPS signal has simplified this type of monitoring. Ship position fixing, conventionally performed using paper nautical charts, has substantially changed, and the ability to continuously verify the ship position has greatly contributed to safe navigation. However, ECDIS cannot yield all the obtainable information.

As explained in the section on the technique of passage planning, the information critical for navigation is written on paper nautical charts during the planning stage. For example, all no go area should be accurately indicated on the chart, because this information is indispensable for safe navigation.

However, entering such critical information is difficult when using ECDIS. Systems such as ECDIS bring numerous benefits;

unfortunately, some functions of maritime techniques that have been developed over a long period of time are not fully compatible with the newly developed instruments.

Next, the use of RADAR/ARPA for position fixing will be discussed. In addition to measuring the distance and direction of a target, RADAR/ARPA is used to continuously ascertain the positional relationship with a landmark through parallel indexing. Occasionally, when instruments that can easily measure the ship position are installed, basic techniques learned for position fixing tend to be either neglected or forgotten. For example, estimating ocean tides and leeway is a function based on position fixing. When position fixing is performed through ECDIS, seafarers may occasionally forget to estimate ocean tides, leeway, or other necessary elements. Seafarers must therefore regularly review the essential maritime techniques to avoid making errors.

3) Communication with other traffic vessels

A critical concern for safe ship navigation is to avoid collision accidents with other traffic vessels. However, under the following circumstances, predicting the future behavior of other traffic vessels presenting risk of collision is difficult:

- If waters are congested with ships or if many ships present risks
- If the navigating area is characterized by the presence of many navigation routes
- If other ships are invisible because of restricted visibility

In all three situations, the future navigation status of the other ships cannot be predicted. When the main navigation route in water areas is defined, and only a single traffic vessel is present, predicting the location and time of the CPA is easy. If the future status can be correctly estimated, then the ship can be handled according to the fundamentals of collision avoidance. However, situations such as those listed occur frequently and require careful monitoring of other ship behavior. Nevertheless, the obtainable information in such situations is limited, which limits the precision of future status predictions.

If the future status cannot be accurately predicted, then communication via VHF wireless telephone is effective and is recommended. In such situations, the following points require attention:

- Timing of communication
- Objectives of communication
- Clear mutual understanding of intentions

Let us consider *communication timing*. After the intention of the other ship is clarified through communication, if the ship must take collision-avoiding action, ensure that adequate time is available to complete the necessary action. Situations must be ascertained and communication established as early as possible.

The *purpose of communication* is for each ship to know the other's intention and then to conclusively decide the appropriate method of realizing safe navigation. Verifying the destination of the other ship and communicating the ship's intentions alone does not meet these objectives. Safe navigation is achieved through mutual understanding.

After communication, actions are initiated on the basis of the information communicated, so a *mutual understanding of the intentions* is essential. A seafarer should not assume that the situation is in their favor. In addition, after communication, whether the other ship is acting according to the information communicated must be verified visually and using RADAR, because one cannot fully ascertain whether the other ship has clearly understood the communicated information.

(3) **Understanding the characteristics of information provided by instruments**

Instruments have characteristic methods of detecting and processing information, and seafarers must therefore have a thorough knowledge of the instruments so that they can select the instrument that would yield the necessary output. In addition, they must be aware of the characteristics of the information provided by each instrument. The precision and reliability of information, any delays in the provision of the obtained information, and other factors must be constantly noted.

(4) **Understanding methods of using provided information**

Observations of behavioral characteristics of inexperienced seafarers have revealed that they often collect more information than necessary but fail to effectively use the obtained information. For example, they may plot all surrounding ships on a RADAR screen and proceed to sequentially look at the ARPA information but would not identify information pertaining to the ship requiring the most attention. In addition, when estimating the ship position through positioning, even if they calculate the degree of deviation from planned course, they occasionally fail to consider whether the deviation is due to wind pressure or tide.

The instruments installed on the bridge provide several types of information. A constant stream of information is available, but not all available information is essential for the task at hand. The required tasks and functions depend on the current navigation situation. Thus, for the necessary functions to be achieved, the provided information must be sorted and appropriately used.

The information provided by the instruments is often limited and must be further processed, following which it can be used in decision making to achieve the various functions. Therefore, it is important that the *functions to be achieved must be thoroughly understood.* For example, during collision-avoidance maneuvers, estimating the distance at which the target ship would pass the bow or stern of the ship and using it as an evaluation criterion is more useful than is CPA in deciding actions. Similarly, TCPA only indicates the time available until collision, and using it to make judgments is not recommended. The necessary collision-avoiding action is decided after considering the speed of both ships and the course crossing angle. The time required to complete the collision-avoiding action differs depending on the course, speed, and encounter conditions of the ship and the other ship. It is incorrect to assume that the collision-avoiding action should be initiated when TCPA is a certain value, for example, 10 minutes. As stated earlier, TCPA indicates the time available for completing the collision-avoiding action, whereas the time required for completing the collision avoidance is determined by the encounter conditions of both ships. Accordingly, the TCPA at which the collision-avoiding action should be taken differs with the encounter conditions. For effective collision avoidance, the necessary information must be collected, integrated, and processed.

Factors affecting the achievement of techniques of instrument operation

(1) Type of available instruments

Seafarers must be aware that the instruments installed on a bridge differ depending on the ship. Similarly, the information obtainable from onshore centers differs depending on the water area. Seafarers must maximize the usable information and apply it to navigation. However, what information is usable depends on the situation, and the available information must therefore be constantly monitored.

(2) Purpose of using information obtained from instruments

The information obtained from the instruments cannot be directly applied in its original form. The collected information should be processed according to its intended use; occasionally, it may be necessary to estimate the missing information. For accurate estimation, it is necessary to understand the characteristics of the obtained information and to correctly understand the way to use it for achieving various functions. Such information processing strongly relates to seafarer competency.

Notes

Here, the author shares notes that have strongly impressed the author during the course of training seafarers in maritime techniques. Regarding RADAR/ARPA, modern ships are equipped with ECDIS, AIS, and other instruments that provide a vast amount of information within a short time. Consequently, seafarers now have access to a large amount of relevant information that is updated from moment to moment. Human information-processing abilities require a fixed time to obtain the information, analyze it, and use the analytical results. In the aforementioned situation, humans can become very occupied while following the updated information and may thus be unable to effectively use the available information. That is, *if a vast amount of information is available, human thinking abilities decrease.* This phenomenon is observed in many inexperienced seafarers. Nevertheless, proficient seafarers do not feel pressured by this information barrage and have the ability to select and analyze only the necessary information. Thus, the author strongly believes that modern navigational instruments inhibit the development of competency in inexperienced seafarers.

KEY FACTORS OF SECTION 2.7: TECHNIQUE OF INSTRUMENT OPERATION

Functions that must be achieved by the technique of instrument operation are as follows:

1) Understand the available instruments.
2) Understand the methods of using instruments to obtain the necessary information.
   The following constitutes necessary information:
   • Information on traffic vessels
   • Information on the ship positioning for safe navigation
3) Understand the characteristics of information provided by the instruments.
4) Understand the methods of using the provided information.

2.8 TECHNIQUE OF HANDLING EMERGENCIES

This technique is for identifying and handling situations of malfunctioning of the main engine, steering gear, and other onboard equipment as well as emergencies outside the ship being handled (hereafter, "the ship") in order to realize safe navigation.

Definition

The technique for identifying emergencies pertaining to the main engine, steering gear, and other onboard equipment as well as those arising in the environment surrounding the ship and for performing the necessary actions to respond to malfunctions and emergencies.

Functions

Appropriate action must be taken during emergencies, for which prior knowledge and training are required. The emergencies that must be anticipated and the corresponding responses must be considered during normal navigation.

(1) **Function for identifying location of problems**

First, consider a case in which an emergency arises onboard the ship. Although problems can occur in any system, problems with systems directly controlling ship navigation, such as the steering system, main engine, and other propulsion systems, are major malfunctions that critically affect safe navigation. The duty officer should confirm beforehand the information to be used for first detection of an arising emergency. Next, they should respond by contacting the engine room regarding the main engine problems to verify the situation. Problems with the steering system lead to an immediate loss of course control, so in addition to collecting information on the bridge, steering functions must be secured through operations such as switching to the emergency steering system and investigating the location of problems. When such situations arise, the seafarers must be competent enough to predict the underlying cause.

In the past, the ability to anticipate and manage various situations was developed by encountering actual situations during long experiences. Currently and in the future, because of shorter onboard periods and the higher reliability of instruments, exposure to actual emergencies is likely to continue to be low, meaning that fewer opportunities are available to develop the appropriate competencies. However, instruments cannot manage all situations, and the danger of emergencies arising is ever present. It is recommended that the management of emergencies should be not only a matter of individual effort but a responsibility shared by society and associations.

(2) **Function for repairing problems and malfunctions**

When a problem arises in the main engine, the duty officer must immediately contact the engine room and request that the cause be identified and eliminated. Problems in steering systems result in an immediate loss of course control, so the steering system must be switched or an emergency steering system activated. Afterward, steps must be taken to repair the malfunctioning steering functions. If necessary, the engine room should be requested to take action.

(3) **Function for restoring necessary functions in the affected parts**

Problematic situations necessitate simultaneous efforts to recover functions and to maintain safe navigation. Emergencies may induce unintended movement of the ship, necessitating contact with and safe distance from nearby traffic vessels. In addition, signal flags may need to be displayed, and traffic control centers or relevant agencies may need to be contacted. Furthermore, if managing the emergency and restoring normal functions is expected to take a long time, the passage plan must be promptly modified, and the new plan communicated to all divisions of the ship.

(4) **Function for identifying and responding to abnormal behavior in other traffic vessels**

When emergency situations arise in ships sailing nearby, seafarers must be prepared to take appropriate action in response to the abnormal behavior of surrounding ships. In addition, even in situations without emergencies, other ships may demonstrate unexpected behavior. In all these situations, a seafarer must constantly ensure safe navigation. For example, when a ship traveling ahead rapidly changes course and makes an abnormal approach, the cause may not always be ship malfunction; the ship may be responding to a dangerous situation around her. Ships may also engage in unexpected behavior without appropriate communication.

Always assume that the other traffic vessels may present a risk to safe navigation of the ship; therefore, always maintain a safe distance from the surrounding ships, and communicate with those ships and traffic control centers to verify the situations when necessary.

Moreover, abnormal behavior in the ship may create a dangerous situation for another ship. Seafarers should ensure that the ship behavior is appropriate from the perspective of the surrounding ships without necessitating communication through VHF radio communication system to verify the maneuvering intentions of the ship.

(5) **Function for identifying and responding to abnormal weather and sea state**

Abnormal weather and sea state mean changes in the natural environment that affect safe navigation. Sudden reduction of visibility, leeway because of unexpected tide, and difficulties in heading control because of strong wind all require appropriate corrective actions. Changes in conditions can be perceived by constantly monitoring the motion of the ship and surrounding ships' circumstances.

If visibility decreases suddenly, measures such as verifying surrounding traffic vessel conditions, checking RADAR/ARPA information, reinforcing lookouts, and using fog signals must be implemented.

Changes in the movement of the ship because of tide, wind pressure, and other external forces must be detected early. Early detection is important to ensure timely risk avoidance planning and execution. This can be accomplished by setting heading targets at the passage planning stage and entering the actual navigating routes on nautical charts as sequential course settings. In addition, parallel indexing is a convenient and effective method in this regard.

When strong winds make it difficult to maintain the planned course, it may become necessary to change course or make preparations for other emergencies. The duty officer must contact the captain for further instructions.

Technique of preparation for emergency situations

The following factors are related to the degree of action achieved in emergency situations such as trouble in the ship, abnormal movement of other traffic vessels, and abnormal weather and sea state. In this section, the necessary preparation for these emergency situations is explained.

Appropriately responding to emergencies requires prior preparation for the types of emergencies. Emergencies normally occur without warning and must be managed within a short period of time. Thus, *advanced preparation* is strongly recommended.

The response measures for emergencies vary depending on the navigational conditions at the time of the emergency. When navigating congested water areas, risks of collision with other traffic vessels must always be considered while selecting the measures. In situations involving restricted navigational waters, when an emergency arises in which control of ship movement is difficult, the priority is to seek measures to prevent grounding. Prior study of the details of various emergencies is recommended.

KEY FACTORS OF SECTION 2.8: TECHNIQUE OF HANDLING EMERGENCIES

Functions that must be achieved by the technique of handling emergencies are as follows:

1) Identify location of problems.

2) Repair problems and malfunctions.

3) Identify and complete necessary tasks related to abnormal occurrences.

4) Detect and react to abnormal behavior in other traffic vessels.

5) Identify and respond to abnormal weather and sea state.

2.9 TECHNIQUE OF MANAGEMENT: MANAGING TECHNIQUES AND TEAM ACTIVITY

This section discusses how to manage techniques appropriately using the eight of the nine elemental techniques mentioned in prior sections in order to realize safe navigation, as well as how to make good use of onboard human resources effectively and efficiently.

Definition

Technical management for combining the eight elemental techniques mentioned in prior sections for safe navigation and team management for maximizing team member competency in order to improve team functions.

Here, the management techniques are broadly divided into technical management and team management.

Technical management

The eight elemental techniques explained in the previous sections are techniques that are independent of each other and are performed to achieve specific functions. From this perspective, each elemental technique is tied to specific functions. Consequently, to achieve actual safe navigation, several individual elemental techniques need to be performed in a necessary combination so that all necessary functions are effectively achieved, as shown in the following specific examples.

(1) **Collision-avoidance maneuvers**

When there is a collision risk, ships presenting a risk are first detected through the *lookout technique* of the abovementioned eight elemental techniques. Then, for effective lookout, it is necessary to collect information on marine traffic vessels through navigational instruments installed on the bridge, such as RADAR/ARPA. How navigational instruments need to be used and what kind of information can be obtained are covered by the *technique of instrument operation*. Techniques concerning the information necessary to avoid ships presenting future risks and the timing of information collection are determined by the necessity to estimate the likelihood of encountering situations in the future, as is needed when making predictions in collision-avoidance situations. From this perspective, the technique of instrument operation can be considered as being performed to improve lookout functions. Thus, multiple techniques need to be performed together in order to effectively achieve each function. Accordingly, determining *how multiple techniques must be combined to achieve the navigational objectives* is termed *technical management*.

Returning to the main topic of collision-avoiding maneuvers, after a collision risk is verified, an appropriate collision-avoidance method must be determined. To verify the future movements of target ships, communication via VHF radio communication system is used. Procedures and English conversation for communication are established as described in the *elemental technique of communication*. However, the communication content and time differ with the situation. *Determining how the communication technique is performed in a particular situation* is an important function of technical management.

COLREG and local rules should be considered in determining collision-avoidance maneuvers. However, even if maneuvers are implemented according to the rules, factors such as ship maneuverability and restrictions on navigable area may restrict the ship operations. In such situations, too, technical management is required, as multiple techniques need to be combined together. In other words, actions are determined through a combination of the *technique of observing laws and regulations, position fixing*, and *maneuvering*. The technique of managing the techniques is indispensable for maintaining safe navigation by applying several independent techniques effectively.

(2) **Measures for emergencies**

Consider a situation in which problems arise in the main engine while navigating a narrow waterway. The *technique of handling emergencies* consists of procedures for detecting problems with the main engine and then returning it to normal operating conditions. However, this function alone cannot maintain safe navigation. Main engine problems reduce rudder effectiveness, which makes heading control difficult. Thus, emergency management entails multiple functions, including verifying surrounding conditions (lookout), contacting the engine room and surrounding ships (communication), modifying passage plans (passage planning), and executing a revised plan (maneuvering).

The important aspect here is to determine the priority of these functions after considering the relationships among functions and to accordingly identify the essential techniques. Thus, *prioritizing and determining the relationship among the necessary functions of each technique is crucial in technical management*.

Team management

Team management pertains to the management techniques needed when multiple seafarers simultaneously contribute to safe navigation and to the realization of team objectives. The technique of team management comprises functions necessary for multiple crewmembers to effectively achieve the team objectives. A leader normally leads a team of other constituent members, referred to as *team members*. For a team to achieve its objectives, the *objectives must be clear* and the *methods for achieving those objectives* must be clarified. The objectives and the methods of achieving them are primarily set by the team leader, who then explains the details thereof to the team members; the tasks that must be accomplished to complete the team objectives are *assigned* by the team leader to individual team members in the corresponding functional units. Therefore, the team leader must *identify the necessary functions for achieving the objectives* and *evaluate the competency of the team members for performing the functions*. Members must perform their assigned functions and then *report the results of tasks* to the leader. Leaders *combine the information reported by each member, analyze the present situation, estimate future situations, and accordingly make and execute an action plan*. Reports by members and commands by leaders are completed mostly orally but occasionally in writing. Thus, communication between team members is an indispensable function for teamwork and is a necessary function of team management. Furthermore, *the functions achieved by each team member as part of the teamwork must contribute to the completion of the team objectives*. If individual team members achieve functions that do not have any relationship with the action of other members and the leader, they will not contribute to teamwork. *The functions achieved by individual members must be closely related. Therefore, members must mutually cooperate in their actions, that is, in achieving their functions. This cooperation is another necessary function of team management.*

The following notes pertain to the application of the aforementioned *functions necessary for team management* to the actions of a bridge team.

(1) Competencies required for team leaders

- The team leader clearly notifies all constituent team members of the details of the objectives to be achieved by the team and must concretely indicate the action required to achieve those objectives.
- The team leader must evaluate the competencies of constituent team members, must issue clear commands regarding the tasks assigned to all constituent team members, and must adequately explain these tasks.
- Team leaders must constantly monitor the behavior of constituent team members, motivate teamwork, and maintain optimum activity. Constituent team member behavior can be judged by the level of assigned tasks achieved and frequency of reporting; so, the team leader must observe them carefully. If team members are found lacking in any function, the team leader must provide appropriate advice or instruction to motivate and correct the team member.
- Team leaders are also members of the team and thus must maintain the following competencies required of team members.

(2) Competencies required for team members

The competencies required of team members listed here are ones that must be attained by team leaders as well as team members.

- Team members must communicate effectively with all constituent team members to share information.
- Team members must constantly engage in cooperative behavior so that the behavior of individuals within the team facilitates smooth teamwork.
- Team members must actively enhance the overall actions of the team and contribute to the achievement of team objectives as laid out by the team leader.
- Team members must fulfill tasks directed by the team leader.

Learning the management technique

The technique of management and its learning differ from the other eight elemental techniques.

(1) Technical Management

Technical management includes important functions such as combining the other elemental techniques and prioritizing and timing the execution of each technique. Technical management is higher level techniques than eight elemental techniques aforementioned The functions of technical management are necessary even though achieving satisfactorily solely by possessing sufficient competency in each of the elemental techniques; further learning is required even after gaining sufficient competency in the individual elemental techniques. Inexperienced seafarers in particular often lack competency in the other eight elemental techniques. Thus, situations often arise in which a lack of competency in technical management may be related to lack of adequate competency in the other eight elemental techniques. Compared with these eight techniques, technical management is an advanced competency. Furthermore, the technique of managing other techniques is essential for normal navigation and is a competency required of even young navigators when on single watch. Therefore, inexperienced seafarers must learn the other eight elemental techniques at an early stage and must continue to learn the technical management technique over time.

(2) Team Management

The technique of *team management* is one part of the management technique and is needed when managing teams of multiple seafarers. The first eight elemental techniques are indispensable for safe navigation when on single watch. That is, they are direct techniques in which information is collected, used, and then implemented to ensure safe ship navigation. By contrast, in situations involving multiple seafarers contributing to ship navigation, the actions of the teams must satisfy the conditions necessary for safe navigation. In other words, *the objective of team management is to ensure that teamwork satisfies the necessary conditions for safe navigation when navigation is performed by a team of seafarers,* as in Part II (Bridge Team Management) of this book. Training in bridge team management is essential to complete the aforementioned objectives.

The management technique, unlike the other eight techniques, has learning prerequisites. For example, *a seafarer can learn technical management only after achieving competency in the other eight techniques.* Furthermore, *compared with technical management, team management requires an even higher competency. In other words, competency in team management can be acquired only after completing the other eight competencies as well as competency in technical management.* There exist examples of seafarers who have received training in team management before acquiring the prerequisite competencies, but this approach improves competency at a very poor rate while also hindering the development of the competencies in the other seafarers on the team. Thus, instructors who teach maritime techniques must clearly understand the characteristics of the individual techniques and accordingly establish a teaching plan.

Functions included in the technique of management

The management technique has mainly been explained in terms of technical management and team management. Seafarers further require competency in many other forms of management to achieve their onboard duties, but these are not discussed in this book. Here, the functions that must be achieved for the technical management and team management, which differ in their respective details but can be generalized as a single function, are described.

(1) **Understanding management targets**

The methods employed differ depending on the management targets, so identifying the targets is important. In technical management, for example, situations that require different techniques have different targets, and the targets determine the details of the technical management. For example, differences in situations may lead to the inverting of the order of priority for techniques.

In team management, the method of management changes with the situations handled by a team and its organizational member structure. The various situations include navigation in narrow waterways, docking, and loading and unloading tasks. The task objectives, role assignments, and standard operating procedures will differ depending on the target situations. Furthermore, they also differ depending on differences among constituent members and the team size. Therefore, seafarers must understand and execute the management methods that change according to these differences.

(2) **Understanding management methods**

Management methods correspond to how each of the described management techniques is performed. Thus, in each situation, seafarers must understand what management actions are to be implemented and how the necessary functions are to be identified.

(3) **Performing the required functions of the management technique**

The functions that must be performed can be identified by understanding the management methods described in the preceding paragraph. The identified functions must then be efficiently performed.

(4) **Evaluating functions**

Understanding the functions that must be performed under a given situation is the most essential factor for determining the necessary management methods, both technical and team management. Anticipating the various situations and identifying the functions needed for each situation, which it is strongly recommended that seafarers attempt, are beneficial for understanding and advanced study of the management functions.

(5) **Evaluating member competencies**

Evaluating member competencies is a requirement of team management. Leaders must assign roles for achieving the necessary functions to constituent team members. To make rational role assignments, the competencies of individual members must be evaluated. Not all constituent members possess the same competency or the same level of competency in various functions. The roles must be assigned such that the team is used optimally.

KEY FACTORS OF SECTION 2.9:
TECHNIQUE OF MANAGEMENT

The targets of management in this book are human resource organization and techniques. The techniques necessary for managing human resource organization are described in Part II (Bridge Team Management). The key factors for technical management are summarized in the following text.

Functions that must be achieved by the management technique are as follows:

1) Select techniques that must be applied.
2) Select concrete functions of techniques to be performed.
3) Determine frequency and timing of performing techniques.
4) Determine priority when multiple techniques must be applied.

In Figure I.2.1, the elemental techniques for management are arranged differently from the other elemental techniques; this is elaborated on in the section on the elemental techniques for management. Figure I.2.2 shows the navigation techniques pertaining to the topmost layer in Figure I.2.1. The key elements of these techniques, which is the primary focus of this book, are indicated.

REFERENCES

Ishibashi A., Kobayashi H.: A Study on the Evaluation of Berthing Maneuver under Wind Disturbance—From Viewpoint of Risk Management (2009), *Proceedings of 9th ACMSSR (Asian Conference on Marine Simulator and Simulation Research)*.

Nishimura T., Kobayashi H.: A Study on Characteristics of a Lookout for Maritime Traffic under Restricted Visibility (2005), *Proceedings of 5th ACMSSR (Asian Conference on Marine Simulator and Simulation Research)*.

Nishimura T., Kobayashi H.: The Relation between Lookout Capacity and Navigational Environment (2009), *Proceedings of 9th ACMSSR (Asian Conference on Marine Simulator and Simulation Research)*.

Chapter 3

Inadequate Knowledge and Competency Often Observed in Inexperienced Seafarers

In this chapter, the results of studies on the behavior of seafarers conducted at the Ship Maneuvering Simulator Center of the Tokyo University of Marine Science and Technology are summarized. In particular, it refers to the research on the behavioral characteristics of inexperienced seafarers conducted to improve the present state of ship navigation, in which the behavior of inexperienced seafarers on actual duty often impedes safety. The behavioral characteristics of inexperienced seafarers reported herein are seen not only in Japanese seafarers but in seafarers from all countries.

This chapter is particularly written for the benefit of inexperienced seafarers and to help experienced seafarers better understand the behavioral characteristics of inexperienced seafarers and thus improve their competency, which is crucial for maintaining the safety of future ship navigation (Uchino and Kobayashi 2009; Ito and Kobayashi 2013).

3.1 CHARACTERISTICS OF INADEQUATE ACTION IN PLANNING

Function of Planning technique by young seafarers with little experience that is preparation before initiating the duty of watch is the main item. Planning consists of the functions described in detail in Chapter 2, which are essential for the preparation tasks. Basic information is entered on nautical charts at the passage planning stage, but seafarers must nevertheless verify all data immediately before starting the watch. Inexperienced seafarers occasionally take insufficient action pertaining to the following points.

(1) Collection of necessary information

Seafarers must always anticipate potential situations immediately before going on duty and while on duty; in addition, they must estimate the predictable situations that will occur during navigation while standing watch. The necessary items to confirm for preparation are altering course plans, navigating the planned course, and communication with marine traffic centers. During navigation situations wherein course alterations are necessary, the bearing and distance to objects as well as the wheel over point must be verified in order to alter the course. The items important for navigating a planned course are the preparations for using parallel indexing and the determination of no-go areas in order to avoid entering risky water areas.

Predictable conditions such as changes in the weather and sea state and their effect on navigation must be verified in advance. In addition, information on encounters with other traffic vessels and risky locations must be collected.

(2) Confirmation of necessary action

The collected necessary information must be implemented as required during navigation. Therefore, the necessary actions determined on the basis of the collected information should be entered onto nautical charts, and preparations must be made for their execution without delay.

KEY FACTORS OF SECTION 3.1: POLICIES FOR IMPROVING THE COMPETENCY OF INEXPERIENCED SEAFARERS; FUNCTIONS ACHIEVED BY TECHNIQUE OF PASSAGE PLANNING

1) To estimate and confirm the situations faced during watch duty before standing watch
2) To enter necessary behaviors for the estimated situations marked on the charts

3.2 CHARACTERISTICS OF INADEQUATE ACTION IN LOOKOUT

The main purpose of lookout functions is to collect and analyze information and to accordingly predict future situations. The characteristics of inadequate action by inexperienced seafarers in each process are as follows.

(1) **Information collection process**

First, regarding the information collection method, little information is collected by visual observation; information is mainly collected using navigational instruments, which enable easy data acquisition. Even when information is collected visually, because the observation range is limited to 30° to the left and right of the bow direction and because the observation time is too short to collect sufficient information, data collection is limited to the detection of surrounding ships. The behavior of other ships must be ascertained and their future movements must be predicted through visual observations.

In many cases, inadequate information collection has led to the delayed detection of ships presenting risks, resulting in the late initiation of collision-avoiding actions and dangerous situations.

(2) **Information analysis process**

Information derived through multiple methods, not obtained from a single source, must be analyzed. The necessary information must be obtained by integrating the information obtained visually and that obtained using navigational instruments. However, frequently, seafarers inadequately compare these two types of information.

The movement of marine traffic vessels is not constant and may change at any time. For example, even when constant movement is maintained, information must be obtained through continuous observation in order to make accurate predictions in the water area. Inexperienced seafarers occasionally do not perform continuous monitoring, do not adequately assess situations, and recognize situations late.

(3) **Prediction of Future Situation**

Without making accurate predictions, one cannot determine which ships present risks and which ones will pass safely. This not only necessitates a sharp lookout for ships that eventually will not require collision-avoiding action but also leads to reduced attention to ships that will necessitate collision-avoiding action, leading to delayed collision-avoidance maneuvers. Improving competency in making predictions in order to identify ships presenting risks and determining the necessity and priority of collision-avoidance maneuvers are crucial for safe navigation.

KEY FACTORS OF SECTION 3.2: POLICIES FOR IMPROVING THE COMPETENCY OF INEXPERIENCED SEAFARERS; FUNCTIONS ACHIEVED BY THE TECHNIQUE OF LOOKOUT

Functions that must be achieved by the lookout technique are as follows:

1) Primarily through visual observation, scan the lookout area both near to and far from the ship being handled.

2) Detect other traffic vessels as early as possible.

3) Collect the following information on the status of traffic vessels:
   - Distance and bearing of other ships to the ship being handled
   - Course and speed of other ships
   - Confirmation of risks to the ship being handled in terms of closest point of approach (CPA), time to CPA (TCPA), and bow crossing range (BCR).

4) Prioritize warnings to other ships on the basis of risks to the ship being handled.

5) Continuously monitor ships presenting risks, initiate VHF communication, and implement collision-avoiding actions in a timely manner where appropriate.

3.3 CHARACTERISTICS OF INADEQUATE ACTION IN POSITION FIXING

Over the recent years, with the development of Electronic Chart Display Information Systems (ECDIS), chart plotters, Global Positioning System (GPS), and other navigational instruments, competency in vision- and RADAR-based position fixing has decreased, which in turn has decreased related competencies in the position-fixing technique as well (e.g., the competency to ascertain distance to an altering course point by using distance to heading objects). This lack of competency to perform position fixing through simple methods means that the position needs to be constantly verified using nautical charts when approaching the point to alter course.

Therefore, during seafarer training at the Ship Maneuvering Simulator Center of the Tokyo University of Marine Science and Technology, the use of ECDIS is deliberately avoided in an attempt to practice RADAR-based position fixing and improve positioning competency. The following are the characteristics of inadequacy in inexperienced seafarers found during training.

(1) **Position-fixing competency**

The use of modern onboard navigational instruments to measure ship position led to RADAR-based measurements requiring a long time; in addition, the precision of such measurements was found to have decreased.

(2) **Use of information obtained from position fixing**

Because of the strong belief that ship position is automatically provided by modern instruments, the inexperienced seafarers had a weak understanding of using the obtained ship position to make predictions regarding present and future status. Estimates of wind and tide conditions, which might affect ship movements, can be made from the ship position obtained through positioning. Careful monitoring of ongoing changes in ship position makes it possible to predict approaches to clearing lines, predict deviations from the planned course line, and determine the appropriate corrective actions. These important functions pertaining to information obtained from ship position measurements cannot be overlooked, and this is a critical competency that must be improved.

In addition, they had insufficient task awareness, including distance and direction to the next course altering point and the calculation of the required speed in order to meet the ETA, meaning that they gave insufficient consideration to some aspects of navigation.

(3) **Relationship of position fixing and other necessary actions**

The timing of position fixing and the other necessary actions was not adequately prioritized. This was often evident in that they persisted with position fixing and updating charts during altering course and collision-avoidance maneuvers.

KEY FACTORS OF SECTION 3.3: POLICIES FOR IMPROVING THE COMPETENCY OF INEXPERIENCED SEAFARERS; FUNCTIONS ACHIEVED BY THE TECHNIQUE OF POSITION FIXING

Functions that must be achieved by the technique of position fixing are as follows:

1) Improvement is needed not only in position fixing using instruments such as ECDIS and GPS but also in competency in position fixing using cross-bearing and radar.
2) The purpose of position fixing is not solely to determine the present position of the ship being handled. The following information must be collected for safe navigation:
    - Measuring direction and distance to next waypoint
    - When a destination has a predetermined ETA, in addition to the direction and distance to the destination, the required speed to realize the ETA
    - Estimations of wind and tide affecting ship movements and their estimated effects on ship movement

3.4 CHARACTERISTICS OF INADEQUATE ACTION IN MANEUVERING

The following points regarding the behavior of inexperienced seafarers were often observed when ships were maneuvered on the basis of the plans.

(1) **Steering commands**

When commands are given to change course because of way-points or when executing collision-avoiding action, two types of commands are used: rudder angle commands and course commands. When making course changes, the rudder angle commands are as follows: steering the rudder to start turning, steering the rudder mid-ship to reduce turning speed when approaching the target heading, and checking the rudder being used for stopping the turning. Because the rudder angle commands must correspond to the situation, seafarers must continuously observe and give commands while changing course. On the other hand, in the case of course commands, the helmsman decides rudder angle and operation timing. Normally, these two types of commands are used according to the turning objectives. However, inexperienced seafarers often do not understand the respective characteristics of these methods, so incorrect steering commands are used to alter the course. Consequently, lookout function and the ability to set the next course are also often insufficient.

(2) **Understanding of maneuverability**

Maneuvering training includes opportunities to learn ship maneuverability before exercise using a simulator. Despite lectures on maneuverability in classrooms and practice with simulators, they have exhibited a lack of understanding of maneuvering, resulting in overshooting and undershooting of the target headings.

In addition, the lack of ability to accurately maintain course or alter course because of the inability to estimate the influence of wind and other external forces has been observed.

(3) **Collision-avoiding action**

The main maneuvers required of inexperienced seafarers are changing course maneuvers at course altering points and altering course to avoid collisions. Inadequate action during altering course maneuvers is as described in the previous section. In this section, competency in collision-avoidance maneuvers is explained as essential for safe navigation.

For collision avoidance, first, the collision risk must be estimated through lookout, and appropriate action must then be initiated. Despite the importance of taking action at the maneuvering stage, inadequate action is often observed. Inaccurate future estimations cause delay to the collision-avoiding action. Moreover, the collision-avoidance angles would be unsuitable, and the resulting repeated small-angle course alterations or excessively large course alterations would lead to approaches to other vessels or extreme deviations from the planned routes, and the objective of ensuring the intended separation from the target vessel would not be realized.

KEY FACTORS OF SECTION 3.4: POLICIES FOR IMPROVING THE COMPETENCY OF INEXPERIENCED SEAFARERS; FUNCTIONS ACHIEVED BY THE TECHNIQUE OF MANEUVERING

Functions that must be achieved by the technique for maneuvering are as follows:

1) Understand the differences in the use of rudder commands and course commands as commands to helmsmen when altering course.
2) Fully understand ship maneuvering performance and issue appropriate steering commands when altering course.
3) Avoid small repeated heading angle changes when making course alterations for collision avoidance.
4) Be careful when making large course changes in order to avoid collisions with a third ship or deviating from the planned course line.

3.5 CHARACTERISTICS OF INADEQUATE ACTION IN OBSERVING LAWS AND REGULATIONS

The minimum code of conduct that must be observed to realize safe navigation is contained in traffic laws. Consequently, seafarers must understand and reflect these laws in their behavior during actual maritime navigation.

However, in navigation, laws and regulations are not reflected in the behavior of inexperienced seafarers, even if they have been learned.

(1) **Articles 5, 6, and 7 of COLREG**

Articles 5, 6, and 7 of COLREG respectively mention "lookout", "safe speed", and "risk of collision". Lookout duties and means are described in Article 5. Articles 6 indicates the appropriate collision-avoiding actions and safe speed considering factors such as visibility, ship maneuverability, traffic conditions, RADAR use, and weather and sea state. The methods of judging a collision risk are described in Article 7. These items are not being fully adhered to by inexperienced seafarers.

(2) **Collision-avoiding action violations within the scope of COLREG**

Even though collision-avoiding action according to laws and regulations is stipulated within COLREG, actions in violation of the rules are often observed. For example, collision-avoiding action is occasionally taken through a series of small course alterations. In addition, occasionally, after overtaking another ship, they would not ensure sufficient space, but set a course that interfered with the course of the overtaken ship.

(3) **Violations outside the scope of COLREG**

COLREG does not clearly define the condition of distance of between a give-away vessel and a stand-on vessel; only the mutual actions between ships within a field of vision are stipulated. Nevertheless, actions are not stipulated for distant ships that are more than 5 miles away. Thus, smooth marine traffic can be realized when ships are navigating more than 5 miles apart; this is not a COLREG stipulation. Inexperienced seafarers have been observed to take actions according to COLREG in relation to distant ships, but this creates new risks or disrupts the traffic flow.

KEY FACTORS OF SECTION 3.5: POLICIES FOR IMPROVING THE COMPETENCY OF INEXPERIENCED SEAFARERS; FUNCTIONS ACHIEVED BY TECHNIQUES FOR OBSERVING LAWS AND REGULATIONS

Functions that must be achieved by techniques for observing laws and regulations are as follows:

1) Understand traffic laws related to navigation and behave in accordance with the laws and regulations.
2) Understand the scope of applicable laws and regulations.
3) Understand laws and regulations applicable to the planned navigation area beforehand, and discuss them with senior seafarers.

3.6 CHARACTERISTICS OF INADEQUATE ACTION IN COMMUNICATION

Communication techniques are classified according to the partners (i.e., other members on the bridge, members in another place outside the bridge, crew on other ships, and marine traffic centers) with whom the information is communicated. Communication with other crew members on the bridge is an important function of bridge team management. In this section, the communication competency for offboard communication partners, which is crucial for safe navigation, is discussed.

(1) **Communication details**

The purpose of communication is the exchange of information that is mutually essential and beneficial. Not fulfilling these objectives often creates problems. For example, inexperienced seafarers may not accomplish the communication objectives as required for the current navigational conditions, such as transmitting the intentions of their own ship and requests and questions. In many cases, before starting communication, they have exhibited a lack of adequate consideration of the necessary information and the logical thinking necessary for achieving objectives.

In addition, even in situations in which necessary information is being acquired from communication partners, only part of the necessary information may be acquired. Before communication, an effort must be made to organize and prepare what is being communicated. The intentions of a partner cannot be understood without sufficient information, and the necessary actions for their ship cannot be appropriately determined.

In cases of inadequate communication, another ship cannot accurately receive the maneuvering intentions of their ship, and a situation is created in which the other ship's intentions cannot be accurately ascertained; this can occasionally result in dangerous situations. Seafarers should remember that they may create risks through their own inadequate communication.

(2) **Communication timing**

When risks are presented by another traffic vessel, mutual verification of intentions and requests for collision-avoiding action are required. However, inexperienced seafarers occasionally make errors in the timing of communication. Occasionally, communication to verify another ship's destination and plan is initiated at the point by which the collision-avoiding action must already have begun. This is due to the inability to recognize the appropriate time at which cooperative action must be initiated. Inadequate timing of information communication is not due to the use of a foreign language but due to a lack of maritime competency necessary for ship navigation. Consequently, to improve communication competency, efforts must be made to improve first, maritime competency and then, the ability to logically comprehend problems.

(3) **Communication language and related factors**

Incorrect and inadequate communication in English often results from a lack of organization of the information communicated. This can be improved by clarifying the objectives of communication and preparing the necessary English expressions beforehand. Inexperienced seafarers tend to take a long time to communicate, but this can be improved by considering the foregoing discussion.

KEY FACTORS OF SECTION 3.6: POLICIES FOR IMPROVING THE COMPETENCY OF INEXPERIENCED SEAFARERS; FUNCTIONS ACHIEVED BY THE TECHNIQUE OF COMMUNICATION

Functions that must be achieved by the technique of communication are as follows:

Communication on the bridge is detailed in Part II of this book. The following are the essential points for communicating with other traffic vessels.

1) Clearly decide the purpose of communication and organize what is being communicated before establishing contact.

2) Ensure that sufficient time will be available for completing collision-avoidance maneuvers after the communication. In particular, communication for collision avoidance should be done as early as possible.

3) Master the basic English phrases essential for maritime communication as early as possible. It is particularly important to organize the issues to communicate in order to engage in a coherent conversation.

3.7 CHARACTERISTICS OF INADEQUATE ACTION IN INSTRUMENT OPERATION

Modern ships have many navigational instruments installed on their bridges, which perform a crucial role in ship navigation. Here, some examples are introduced, which show that the use of instruments could significantly affect achieving safe navigation. Observations of inexperienced seafarers from the perspective of whether they had adequate knowledge of and were effectively using the instruments have revealed the following problems.

In a study of seafarer behavioral characteristics conducted by the Ship Maneuvering Simulator Center of the Tokyo University of Marine Science and Technology, basic characteristics of information collection and maneuvering were studied in navigational situations. Seafarers executed ship handling without ECDIS or Automatic Identification Systems (AIS) support, which revealed the behavioral characteristics of instrument operation under such conditions. By establishing observable conditions for clear estimates of the behavioral characteristics of seafarers, the information and methods used by the seafarers were monitored.

(1) **Use of RADAR/Automatic Radar Plotting Aids (ARPA)**

The following inadequate actions were observed in conditions in which RADAR/ARPA installed for collecting information (e.g., for positioning or observing other ships) was used.

1) Selection of RADAR range

RADAR range corresponds to the lookout range, and the detection of long-range targets, including those that present risks, is delayed if a short RADAR range is used. In addition, if a long range is set, then small objects and ships in the vicinity of the ship being handled cannot be detected. During poor visibility, RADAR range tends to be limited to short ranges, which results in late detection of long-range ships presenting risks. Efforts should be made to constantly switch the RADAR range in order to monitor long- and short-range distances.

Furthermore, if RADAR is used for positioning, then targets within the shortest available RADAR range are used, and the targets are set close to the outer ring, which is the recommended approach.

2) Selection of target vector length

Motion vectors provide useful information for estimating the movement of other traffic vessels and future risks. However, if motion vectors are set small in long-distance RADAR mode, then the vector length will be short, and determining vector direction and detecting course changes will be difficult.

3) Use of basic RADAR functions

RADAR has secondary functions, such as setting an electronic bearing line, but the use of these functions to ascertain risks from other ships or to realize safe navigation through parallel indexing has rarely been observed among trainees. RADAR functions are not often used to estimate whether other ships will pass the bow or stern of the ship being handled or to estimate the passing distance.

4) Use of ARPA information

ARPA displays numerical data pertaining to other ships and can thus assist in accurate decision making. However, the number of ships that can be monitored simultaneously is limited, so the targets must be appropriately selected. Inexperienced seafarers frequently do not select ships that require the most attention or collect the most appropriate information.

As discussed earlier, instrument handling is closely related to the level of ship handling competency. The information obtained from instrument operation must be appropriately interpreted, and the methods of using it must be fully understood.

(2) **Use of head-up display and related features**

Various instruments other than RADAR/ARPA are installed on the bridge of ship handling simulators at the Ship Maneuvering Simulator Center. These include instruments showing the operation of main engines and their operational status, instruments related to steering systems, whistles and other signaling devices, ship speed indicators, propeller revolution indicators, wind speed and direction indicators, rudder angle indicators, and rate-of-turn meters, all of which are important for navigation and require constant monitoring. These are installed over the ship's front window as a head-up display. These instruments are often not monitored; thus, status assessments and predictions are poor, and little attention is paid to the turning angle and changes in ship speed.

KEY FACTORS OF SECTION 3.7: POLICIES FOR IMPROVING THE COMPETENCY OF INEXPERIENCED SEAFARERS; FUNCTIONS ACHIEVED BY THE TECHNIQUE OF INSTRUMENT OPERATION

Functions that must be achieved by the technique of instrument operation are as follows.

The purpose of using navigation instruments is to obtain effective information for navigation. Seafarers must have a strong intention to use the information for safe navigation.

1) When using RADAR/ARPA:
   - Timely switching of radar range and observation of both long- and short-range targets
   - Setting motion vector lengths of other ships corresponding to the radar range
   - Using parallel indexing and BCR information and applying it to maneuvering.
   - Attaining competency in prioritizing ships presenting risks and collecting the necessary information.
2) Organizing information obtained from instruments installed on the bridge, understanding how to utilize it, and applying it to navigation.

3.8 CHARACTERISTICS OF INADEQUATE ACTION IN MANAGEMENT

In general, onboard management covers a very wide variety of areas, but the first requirement of young seafarers is the appropriate use of techniques necessary to maintain safe navigation. To this end, they must know the functions achieved by these techniques and execute them effectively when appropriate. Accordingly, the technique of managing these functions is generally defined as technical management. In this section, the results of a study of the behavioral characteristics of inexperienced seafarers from the perspective of technical management are summarized.

(1) **Managing techniques**

The eight elemental techniques other than the management technique must be appropriately performed in a timely manner. For example, on encountering a ship with a collision risk, several elemental techniques must be performed, including lookout, communication, instrument, observing laws and regulations, and maneuvering. In addition, they must be prioritized and timed appropriately. The procedure for performing the techniques varies with the situation, and the appropriate techniques must be selected and performed at the correct times. However, inexperienced seafarers often cannot do so because they cannot determine what functions are necessary or they cannot recognize changing situations.

(2) **Selecting techniques**

Situations often arise in ship navigation in which multiple elemental techniques must be performed almost simultaneously. For example, occasionally, lookout, positioning, and communication must be performed within a short period. In such cases, the importance of each of the functions must be determined and the functions accordingly prioritized. Inexperienced seafarers are weak in appropriate decision making on prioritization based on the current circumstances.

Inexperienced seafarers are often inadequate in achieving basic elemental techniques, particularly technical management. This is because technical management requires a complete understanding of the functions of each of the elemental techniques and the methods to perform them. Consequently, high competency in the first eight elemental techniques is essential before seafarers can acquire adequate competency of technical management. Thus, inexperienced seafarers must make an effort to first master each of the basic elemental techniques.

KEY FACTORS OF SECTION 3.8: POLICIES FOR IMPROVING THE COMPETENCY OF INEXPERIENCED SEAFARERS; FUNCTIONS ACHIEVED BY THE TECHNIQUE OF MANAGEMENT

Functions that must be achieved by the technique of management are as follows.

An example of insufficient competency seen in inexperienced seafarers is competency in technical management. Competency in technical management requires the following:

1) Proper selection and timely performance of techniques: it is necessary to know when and how to perform techniques in navigational situations requiring the execution of multiple techniques.
2) When multiple elemental techniques must be performed simultaneously, the necessity and priority of the functions must be established.

The technique of management on techniques cannot be achieved without fully understanding the function of each elemental technique and how to perform it. Therefore, competency in each of the elemental techniques must first be improved.

REFERENCES

Ito K., Kobayashi H.: A study on the Inexperienced Mariner's Behavior Characteristics Based on the Analysis of Maritime Techniques Regarding Ship Handling (2013), *Proceedings of 13th ACMSSR (Asian Conference on Marine Safety and System Research)*.

Uchino A., Kobayashi H.: Characteristics of mariners' experience and influence on an accident (2009), *Proceedings of 9th ACMSSR (Asian Conference on Marine Simulator and Simulation Research)*.

Chapter 4

Significance and Use of Elemental Technique Development

4.1 SIGNIFICANCE OF ELEMENTAL TECHNIQUE DEVELOPMENT

In Chapter 2, it is explained that many of the actions performed onboard are techniques for achieving safe navigation. Thus, many onboard actions correspond to certain necessary techniques, and these techniques were categorized into nine groups according to the intended functions. Because these nine basic techniques are mutually independent from each other in function, they are termed *elemental techniques*. The various situations encountered during navigation can be adequately addressed by combining the necessary elemental techniques for achieving safe navigation. The significance of the categorization is now discussed.

(1) **Safe navigation achievement and necessary techniques**

The requirements for realizing safe ship navigation are explained in Section 1.4 of Chapter 1 "Conditions Necessary for Safe Navigation". Figure I.1.8 from that section is reprinted here as Figure I.4.1.

The vertical axis in the figure indicates the achievable competency level of seafarers. The horizontal axis indicates the necessary competency level required by the navigational environment. The relationship between the two can be interpreted as follows. Let us imagine a condition of navigational environment, such as a coastal navigation by a very large crude carrier (VLCC). If we also consider the tide in this water area, adequate measures need to be taken against grounding accidents. In such an environmental condition, ship position needs to be frequently measured and shallow water area perceived prior to approach. A necessary condition is the frequent high-precision measurement of ship position. This water area can be considered to be one requiring measurements every 5 minutes within an error of 0.5 cables. This refers to the situation shown on the horizontal axis. When it is possible to measure ship position every 5 minutes within an error of 0.5 cables, the achievable technical level demonstrated by the seafarers navigating this ocean area is equal to the level required by the

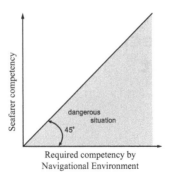

Figure I.4.1 Necessary conditions for safe navigation

environment, and ship position measurements for safe navigation are satisfied. On the figure, this situation appears as a point on a 45° line. This 45° line represents the situation in which the respective competency levels shown by the vertical and horizontal axes are equal.

In this situation, ship position measurements can also be made with a seafarer technical level that is within an error of 0.5 cables every 3 minutes. In this case, the achievable seafarer competency level is higher than the level required by the environment. This state is represented by a point within the area above the 45° straight line on the figure. Since the achievable competency level of seafarers is higher than the level required by the environment, ship position measurements necessary for safe navigation are satisfied, and so are the conditions for safe navigation.

By contrast, in cases where the achievable competency level of seafarers is low and ship position measurements are not possible every 5 minutes within an error of 0.5 cables, the conditions of safe navigation are not satisfied, resulting in a dangerous state. On the figure, this state is represented as a point within the area under the 45° straight line, which indicates dangerous navigational conditions.

Thus, for safe navigation to be realized, realizing the necessary techniques to the level determined by the environmental conditions is an indispensable requirement. The environment in the example only required position fixing, but typically, the environmental conditions usually require various and multiple techniques. Even in such cases, conditions on the horizontal axis are decided by what elemental techniques are required and to what level. Thus, navigational safety can be estimated by estimating the achievable level of competency versus the level required by the environment. Ship navigational safety can be estimated by balancing these parameters.

(2) **Clarification of necessary techniques in target situations**

What action seafarers must take to achieve safe navigation in the various situations faced onboard is important.

For example, consider the situation in which another traffic vessel is encountered during navigation. The following actions must be performed by the seafarers to avoid colliding with the vessel:

1) Detection of the other vessel
2) Estimation of risks to the ship being handled (hereafter, "the ship") from the other vessel
3) Determination of means of avoiding collisions
4) Execution of collision-avoiding actions
5) Verification of the effects of the avoidance action

The collision-avoiding action discussed here corresponds to the elemental techniques required at each stage. First, the traffic vessels must be detected. The relevant technique at this stage is lookout, and the instrument operation technique is also required depending on the lookout method. Next, future impending risks to the ship must be estimated. This is achieved through continuous lookout; RADAR/ Automatic Radar Plotting Aids (ARPA) and other instruments may also be used. When lookout and instrument operation competency are inadequate at this stage, predictions may be late or mistaken, and dangerous conditions may arise. If accurate predictions cannot be obtained, then very high frequency (VHF) or other communication instruments may be used to contact the target ship and verify her intentions. This is a situation in which the communication technique is applicable. When a future collision risk is verified, then preparations are made for collision-avoiding action. Collision-avoidance methods are determined by observing the traffic laws and in accordance with a plan based on understanding the maneuverability of the ship and lookout for the surrounding ships. After a collision-avoidance plan is finalized, the changes in the situation are monitored through the lookout technique, and the starting time for collision-avoiding action is decided through technical management. Subsequently, the plans are executed through the maneuvering technique. As collision-avoiding action proceeds, its effects are verified by lookout, by monitoring the change in the closest approach distance and the encountering situation, and by monitoring whether safe passing is ensured. Since collision-avoiding action results in deviation from the planned course, ship position must be verified through position fixing.

Thus, the eight elemental techniques are applied at the appropriate times, even when it is only a matter of following collision-avoidance procedures. In addition, to simplify the discussion, only a sampling of the applicable techniques has been mentioned here; a more detailed analysis would reveal the further involvement of the technical management, communication, and instrument operation techniques at each stage. Furthermore, in the abovementioned situation, the occurrence of an emergency condition between the ship and another ship has not been assumed, so no additional emergency treatment technique situations are discussed.

Actual navigation situations (for example, navigating congested water areas under low-visibility conditions, harbor navigation, and docking) can entail the synthesizing of all nine elemental techniques. Thus, necessary elemental techniques can be clarified for each maneuvering situation.

KEY FACTORS OF SECTION 4.1: SIGNIFICANCE OF ELEMENTAL TECHNIQUE DEVELOPMENT

The following are made possible by developing elemental techniques for ship handling:

1) By analyzing safety in ship navigation based upon necessary techniques, it becomes clear that the attainment of necessary techniques determined by environmental conditions is a necessary condition for realizing safe navigation.

2) The techniques necessary for safe navigation in any situation can be organized into nine technical elements, and safe navigation can be achieved by synthesizing these nine elemental techniques.

4.2 TECHNIQUES AND COMPETENCY

Researchers have analyzed the tasks performed by humans within systems; most of these studies were conducted from the perspective of information processing or control in human–machine systems. However, no studies have addressed these interactions from the perspective of human beings as the center of all systems, and nor have studies analyzed these systems from the perspective of the necessary functions of human beings. The analysis of techniques performed by human beings introduced in this book is the result of analyzing seafarer functions and clarifies the functions that must be achieved by humans. This analysis clarifies various aspects that were unclear in the past. In this book, all human activities and meanings in system are discussed considering seafarers as the agents. In a system composed of ships and environments, the implementation of the necessary techniques is understood as the ability of seafarers to accomplish these techniques. Seafarer competency can be discussed from the perspective of elemental technique development.

(1) **Clarifying the abilities for performing navigation techniques as seafarer competency**

Seafarers encounter various maneuvering situations while onboard, and it is their duty to manage these situations appropriately to achieve safe navigation. To perform this duty, seafarers *require the ability to achieve the necessary techniques*. Thus, this ability is defined as *the maritime competency that seafarers must master*. Seafarer competency is judged by the adequate performance of the elemental techniques necessary in each situation. The functions that must be achieved through the action of seafarers are determined through elemental technique development. Seafarer competency can be clearly judged by comparing the functions required of seafarers for safe navigation and those achievable by the seafarers.

(2) **Use for technical training**

If the technical proficiency demanded of seafarers is defined, then the target techniques in training can be clearly determined. Conventionally, training is provided to improve seafarer competency in maritime techniques under various maneuvering situations. However, in this approach, it is unclear which specific competencies are being taught, and occasionally, the same competency is being taught for different maneuvering situations. If the competencies taught during training are categorized by maneuvering situations, then a wide range of training situations is required for covering all the required maneuvering situations. Thus, developing the core competencies required for actual navigation would require a plethora of training courses.

By contrast, a logical training course can be developed by considering the relationship of the techniques covered in training to the elemental techniques. The competency level of the same elemental techniques may differ depending on the maneuvering situation. If the details of one elemental technique are analyzed, then the level of competency improvement in training can be clearly ascertained.

Note that the conventional training systems need not be completely changed. A more logical training setup can be established based on the current setup by reviewing the techniques targeted for competency improvement against the elemental techniques. Consequently, the content offered in all currently available training courses would be recategorized under the necessary elemental techniques.

For example, if training is provided for collision avoidance in open water areas with good visibility, the target techniques for training are as follows:

1) Lookout technique (emphasis on vision-based lookout)
2) Maneuvering competency for avoiding collisions
3) Techniques observing traffic regulations for collision prevention
4) Technique of technical management

If the same training must be provided but under poor visibility conditions, the target techniques for training are as follows:

1) Lookout technique (emphasis on instrument-based lookout)
2) Maneuvering competency for avoiding collisions
3) Technique of observing traffic regulations for collision prevention
4) Technique of instrument operation
5) Technique of technical management

Even in these simple examples, the training may involve overlapping elemental techniques. Given that training must be efficient and effective, the techniques targeted for training must be clarified, and a logical training system must be created.

New instruments are frequently installed during actual ship navigation. The techniques to handle these instruments must be fully understood. However, instrument handling training is focused only on the handling of the instrument itself, which is inadequate from the perspective of actual use. Learning how to use the instruments in actual navigation situations is important, and installing these new instruments at the training facilities would therefore be valuable. For example, in the aforementioned two examples, if visibility is less than 5 miles, the training can emphasize the use of information obtained from instruments in addition to that obtained from visual lookout in the visible area.

(3) Use for assessment of competency

If the functions that seafarers must achieve are identified as the nine necessary competencies, then it becomes possible to set the techniques required for each maneuvering situation and the necessary competency level for performing each of these techniques. That is, seafarers who are required to achieve necessary functions at a necessary achievement level in certain situations are required to perform the techniques corresponding to those functions at the required level. The ability to accomplish a technique is defined as competency in that technique. By clarifying the competency that must be achieved in each technique, standards for the competencies required of seafarers can be established. Subsequently, seafarer competency can be evaluated by comparing it against the required competency standards.

In general, the level of competency required varies with the maneuvering situation, and so does the competency level a seafarer must possess to manage that situation. Table I.4.1 shows the relationship between each elemental technique and different maneuvering situations.

This table shows the competency that must be acquired through training and thus indicates the elemental techniques and standard competency that must be acquired by the respective seafarers and seafarer candidates. As shown in the table, seafarers holding a high license rank would have at least the same competency required of seafarers holding a lower license rank for the indicated function; that is, a junior officer mark in the table indicates that a seafarer with a higher license rank (i.e., a senior or a captain) also has that competency. In addition, cells with two marks indicate that the necessary techniques change depending on the environmental conditions. For example, maneuvering competency in anchoring when the environmental conditions are calm is the maneuvering

Table I.4.1 Relationship between the elemental techniques for ship handling and the navigation situations (i.e., seafarer license rank for each function)

	Ocean Operation	Coastal Operation	Narrow Channel Operation	Entering/ Leaving Port	Collision Avoidance	Anchoring	Berthing
Planning			□ ●	○		□ ●	●
Lookout	○	○	○	○	○ □	○	○
Position Fixing				□		□	
Maneuvering	○	○	□ ●	□	○ □	□ ●	●
Observing Laws and Regulations	○	○	□	○ □	○ □		
Communication	○	○	□	□	○ □	□	□ ●
Instrument Operation	○	○	□		○ □	□	□
Emergency Treatment		□	□ ●	□ ●		□ ●	□ ●
Management		○	□ ●	□ ●	○ □	□ ●	□ ●

○: For Cadets and Junior Officers; □: for Senior Officers; ●: for Captains and Pilots

competency achievable by a senior officer, but when the environ-mental conditions are severe, maneuvering would require the com-petency of a captain.

KEY FACTORS OF SECTION 4.2: TECHNIQUES AND COMPETENCY

The following are made possible by clarifying seafarer competency:

1) By comparing the required competency level in a function against the actual performable function level of seafarers, seafarer competency can be clearly evaluated.
2) By clarifying the relevant techniques of implementing efficient and effective competency training, a logical training system can be created.
3) The required competency level differs depending on the navigation situation. Seafaring qualification ranks are determined according to the required competency level. Thus, the difficulty of a navigational condi-tion is evaluated according to the competency level required to ensure safe navigation in that environment.

4.3 LIMITS OF SEAFARER COMPETENCY
AND COMPETENCY EXPANSION

A valid consideration is that there are limits to seafarers' competency when they execute a technique. For example, there are limits on the frequency and accuracy of measurements to achieve position fixing. It is illogical to think that position fixing with unlimited frequency and 100% accuracy is possible. It needs to be understood that there are limits to the extent to which these techniques can be accomplished; that is, there are limits to competency. A variety of reasonable considerations can be made from this viewpoint. In the following, some typical examples are provided.

(1) **Identifying causes clearly during accident investigation**

Although the causes of various accidents have been analyzed and new rules created, navigation accidents have not decreased, and although 80%–85% of accidents are reportedly due to human decision or action, few studies have analyzed why seafarers take actions or are forced to engage in actions that lead to accidents. If we assume that the cause of most accidents is human behavior, then measures to reduce accidents would be impossible without studying human behavioral characteristics. Therefore, the limits of competency must be known.

(2) **Use in accident cause analysis**

Analyses of accidents involving ships have revealed that numerous accidents are caused by the inadequate achievement of seafarer functions. Seafarers are responsible for many functions when navigating ships. By analyzing the process by which an accident occurs, the stage at which inappropriate decisions are made and action taken can be identified. The inevitable result of assigning so many functions to seafarers is that seafarer behavior is the cause of most accidents.

If the objective of cause analysis is to prevent accident recurrence, then the causes of the inappropriate behavior in seafarers must be analyzed. In particular, the elemental techniques related to the inappropriate behavior and the necessary conditions to achieve these elemental techniques must be investigated. Routine accident cause analysis *typically focuses on identifying the techniques corresponding to inappropriate seafarer behavior, and not on why the inappropriate behavior occurred*. Thus, accidents are attributed to seafarers, and there is no further investigation into the reasons for the inappropriate behavior. This lack of true cause analysis results in inadequate corrective measures and thus a lack of reduction in the number of accidents.

As is clear from the foregoing discussion, the reasons why seafarers engage in inappropriate actions must be found in order to clarify the true causes of accidents. Furthermore, the necessary conditions for realizing the appropriate action need to be recognized. Clarifying these conditions, which would lead to the achievement of the individual elemental techniques, requires the analysis of seafarer behavioral characteristics. This is both fundamental and essential. A few researchers are currently engaged in this topic.

(3) **Developing navigational instruments**

Seafarers have a duty to achieve the necessary functions as determined by the maneuvering conditions. However, when we examine the characteristics of seafarers who achieve functions, we find there are limits to behavioral characteristics and the achievement of functions.

For example, consider lookout functions. During narrow visibility conditions, lookout is mainly through RADAR/ARPA, in which the lookout range is closer to the ship being handled (hereafter, "the ship") than in lookout during good visibility. In addition, when surrounded by multiple ships presenting risks, even under good visibility, only a limited number of ships can be observed carefully. These negative aspects of seafarer behavioral characteristics are manifestations of *human limitations on achieving functions*. In either case, from the perspective of achieving necessary functions, *standard seafarer behavior and limits on achieving functions are negative factors. Improving these negative factors is the original motivation for developing the navigational instruments.*

As peripheral technology has developed, new navigation instruments of various types and with various functions have been developed and realized. However, most of this development is "seeds oriented" and not "needs oriented": instruments are developed because the technology for the development exists and not because there are functions required of seafarers. There are several reasons for this discrepancy. One is that the distance between the place where the instruments are used and the place where they are developed is quite large. This corresponds to the general view that good instrument development is difficult when both parties are far apart. A second cause is that engineers developing instruments do not really understand the functions of seafarers. *It is therefore hoped that this book will be read by developers of navigational instruments in order to understand seafarer functions* (Ishibashi and Kobayashi 2006).

Navigational instruments can be categorized as instruments accepted by seafarers as useful and those that are not used even when installed. For an instrument to be deemed useful, it must provide or realize the information or functions that seafarers need. Accordingly, how instruments contribute to the techniques that must be performed by seafarers must be understood by the instrument developers.

(4) **Establishing navigational environment**

The achievement of necessary elemental techniques to realize safe navigation by seafarers has already been described in this book.

Moreover, the necessary conditions must be satisfied for the functions to be achieved. This is due to seafarer behavioral characteristics and the limitations on function achievement, as described earlier. For example, by setting a fairway, the maritime traffic flow becomes uniform, and random traffic flow becomes reduced, which reduces the lookout burden. Thus, lookout achievement remains within human capability, and seafarers attain sufficient lookout for safe navigation.

In addition, when the traffic information is provided, the present status and future intended behavior of other traffic vessels is clarified. It thus becomes possible to determine the current behavior of the ship on the basis of long-term predictions. Compared with maneuvering based on navigational planning according to long-term predictions, maneuvering based on short-term predictions is more dangerous.

(5) **Estimating safety of the navigational environment**

Seafarers show limitations in the achievement of functions. Therefore, an environment must be maintained in which safe navigation can be realized within the levels of functions that seafarers can achieve. In situations in which the necessary functions for safe navigation exceed the human capacity for realizing these functions, groundings, ship–ship collisions, and other accidents may occur. The seafarer competencies related to these accidents are related to each of the elemental techniques. By assessing whether the environment is one in which each of the elemental techniques is fully achievable, *the safety of the navigational environment can be estimated.*

KEY FACTORS OF SECTION 4.3: COMPETENCY LIMITS AND COMPETENCY EXPANSION

The following points arise if it is assumed that seafarer competency is limited:

1) Seafarers have limitations in the level of their competency to achieve necessary techniques.
2) The limits of seafarer competency must be known in order to find the real causes of accidents. In addition, the conditions for attaining competency must be known.
3) Navigation instruments become effective when they enhance seafarer competency to achieve functions.
4) The development of the navigational environment must be based on the ability of the seafarer to achieve the necessary functions.
5) The safety of a navigational environment can be quantified by using standard seafarer competency as a measure.

REFERENCE

Ishibashi A., Kobayashi H.: A Study on Development and Evaluation of the Ship Handling Support System (2006), *Proceedings of 6th ACMSSR (Asian Conference on Marine Simulator and Simulation Research).*

Part I

Postscript

This book is a collection of the author's experiences from more than 20 years of competency training of working seafarers and the findings of analytical research on navigational techniques.

In Part I, the experience of MET (Maritime Education and Training) and the results of their research, a comparative study on seafarer behavioral characteristics during the course of training as well as research on techniques for safe navigation, are summarized. The results have been presented to several international academic societies. This book collates these research results to develop a system of navigation techniques.

The author is a graduate of Tokyo University of Mercantile Marine, where he has also been involved in research and education as a professor. However, neither the necessary techniques for ship navigation nor explanations systematically covering their study were available. Broadly speaking, the techniques were expressed using the word "seamanship". Although breaking this down into specific details yields representative techniques such as lookout, maneuvering, and observing regulations, they have not been organized in a general form that encompasses all the necessary techniques. When a system is developed as a construct covering all the techniques for safe navigation, it clarifies the necessity and relative importance of each technique in a situation. Thus, the conditions necessary for safe navigation are identified. Such a system would serve as a system for learning, which can then be further developed through academic study.

In this book, a system of navigation techniques is described from the aforementioned perspective. Furthermore, to understand the function of each of the elemental techniques constituting the system, how these functions correspond to each elemental technique is explained in relation to the realization of safe navigation. As mentioned in the preface to Part I, this book does not attempt to explain practical actions or knowledge for acquiring each elemental technique. The author would prefer that such detailed knowledge was learned separately.

However, many references on elemental techniques have explanations that are often disconnected from actual practice. Thus, in this book, each elemental technique has been described in relation to the actual tasks for safe navigation. In particular, inadequate competency and behavioral characteristics seen in inexperienced seafarers serve as a useful reference for actual work.

Part II

Bridge Team Management

Chapter 5 Techniques Necessary for Safe Navigation 115
Chapter 6 Factors in Achieving Safe Navigation 121
Chapter 7 Background of Bridge Team Management 131
Chapter 8 Bridge Team Management 147

Part II

Preface

Techniques for navigating ships were developed very early in the long course of human history. These techniques have gradually accumulated and advanced with human cultural growth. The author has learned and researched contemporary ship handling techniques, although he does not have detailed knowledge on the history of ship navigation techniques. Through this research process, the author has come to understand the overall picture. Presently, techniques for ship navigation are considered an aggregate of related techniques developed over time at places where one can learn navigation techniques. These techniques are mentioned in the International Convention on Standards of Training, Certification and Watchkeeping for Seafarers of the International Maritime Organization. To clarify the details of the techniques for safe navigation required of seafarers, the actual techniques employed must be logically organized in terms of their functional aspects.

In contemporary society, many forms of technical knowledge contribute to human activities. Technical systems contributing to contemporary society, such as shipbuilding and architecture, have been established through the organization of the various elemental studies required to achieve specialty techniques in each field. Shipbuilding engineers, for example, learn the various theories needed for shipbuilding and then employ these as the techniques needed for constructing actual ships.

Similarly, ship handling entails various techniques. However, the establishment of these theories has not been completed, and the actual use of the techniques is still developing. Thus, *techniques tend to be treated as an aggregate and have not yet been organized into a system*. If techniques were systematically organized, then various techniques could be sorted in terms of their mutual independence, following which the relationship among the mutual functions of these techniques could be clarified. Subsequently, how each of these functions contributes to achieving the objectives of the system could be defined. By contrast, when techniques are treated as an aggregate, only the general meaning can be understood, and a detailed discussion of individual techniques becomes difficult.

Modern science has advanced through the analysis of research subjects. Through analysis, the constituent elements of a phenomenon can be identified, following which the meaning and content of the functions performed by these elements in order to accomplish the phenomenon are clarified.

In Part II of the book, techniques related to ship navigation, particularly bridge team management (BTM) and bridge resource management (BRM), are analyzed and explained from the aforementioned perspective. The meanings of these words are explained in Part II. In addition, the author would like to point out that *this book has been compiled for the purpose of explaining these words based on scientific analysis, which intends to prevent their convenient use in society without sufficient understanding of their meaning.*

What makes this book different from previously published manuals on navigation and seamanship is that it is a technical manual based on scientific analyses, which are included here. Furthermore, let the author remind readers that the effectiveness of the concepts explained in this book has been proven by the results of training many seafarers.

In Chapters 5 and 6 of Part II, the conditions necessary to achieve safe navigation are explained. Some parts overlap with Part I of this book. Readers who start reading from Part II must first understand *the conditions necessary for safe navigation*, which are the foundation of BTM, so the relevant sections have been summarized in Chapter 1.

Chapter 5

Techniques Necessary for Safe Navigation

The necessary techniques for safe navigation are described in the International Convention on Standards of Training, Certification and Watchkeeping for Seafarers (hereinafter, STCW) of the International Maritime Organization (IMO). The techniques included in the STCW range from those necessary for safe navigation to those necessary when working in teams. This book focuses on ship handling techniques by plural seafarers and the necessity of bridge team management (BTM). BTM is intended to ensure safety. Therefore, first, the conditions necessary for realizing safe navigation are discussed.

Ships complete safe navigation without accidents by employing various techniques onboard. Ships encounter many different situations over a voyage, and seafarers must constantly make practical decisions to execute the techniques appropriately. The techniques they must employ in these encounters are many and various. Therefore, if we analyze the actions seafarers take one by one, it becomes clear that the techniques they use can be categorized on the basis of several factors. Namely, the techniques for realizing safe navigation can be considered an aggregate of several independent techniques that do not overlap. The results of extensive analytical research on the organization of the necessary techniques according to various navigation situations have indicated that the techniques can be broadly categorized into the following nine elemental techniques:

1) Passage planning: Technique of making a plan for navigation
2) Lookout: Technique of observing and estimating other traffic vessels and conditions in the surrounding environment
3) Position fixing: Technique of estimating the ship position
4) Maneuvering: Technique of controlling the movements of the ship being handled
5) Observing laws and regulations: Technique of understanding and following the rules of navigation for marine traffic
6) Communication: Technique of communicating inside and outside the ship being handled

Table II.5.1 Definitions, functions, and factors influencing the achievement of functions of the nine elemental techniques

Elemental techniques	Details
1. Planning (Definition)	Techniques of collecting information on the navigational environment, creating passage plans, and creating plans to execute these plans.
(Main functions)	(1) Understand the information necessary to create plans (2) Understand the methods of using the necessary information (3) Apply planning information to the actual plans (4) During navigation, make changes to plans if the situations are different from those predicted during the initial planning
(Influencing factors)	(1) Traffic rules and regulations (2) Quality and quantity of effective information for navigation (e.g., presence of the recommended fairway) (3) Quality and quantity of obtainable basic information (Information on weather, seas, geographical features, waters, and navigation) (4) Navigation area (ocean navigation, coastal navigation, narrow waterway navigation, fairway navigation, harbor navigation, river navigation) (5) Purpose of navigation (navigation at sea, dropping anchor, docking)
2. Lookout (Definition)	Techniques of detecting stationary targets and moving targets; identifying them; estimating the type, distance, direction, moving speed, and moving direction of targets; and predicting future risks.
(Main functions)	(1) Identify the present situation (types of ships encountered and position and movement of target ships [i.e., course, speed]). (2) Predict the future situation: movement of target ships (i.e., future position, course, speed), changes in the movement of targets, estimated risks to own ship (closest point of approach [CPA], time to CPA [TCPA], bow crossing range [range from bow or stern: BCR]).
(Influencing factors)	(1) Navigational instruments (compass, RADAR, RADAR/ARPA, AIS, vessel traffic information service [VTIS] information) (2) Visibility (3) Volume and flow characteristics of marine traffic vessels (4) Navigational conditions (in oceans and fairways) and traffic laws
3. Position fixing (Definition)	Techniques of estimating own-ship position by selecting optimal objects visually and by using navigational instruments. Techniques of estimating factors affecting own-ship movements and their magnitude.

(Continued)

Table II.5.1 (Continued) Definitions, functions, and factors influencing the achievement of functions of the nine elemental techniques

Elemental techniques	Details
(Main functions)	(1) Select methods of collecting information for position fixing (select measuring instruments; select objects for position fixing) (2) Estimate own-ship position (achieve required accuracy and frequency). (3) Estimate own-ship movement status (estimate direction of movement, speed of movement, rate of turn, wind, and tide)
(Influencing factors)	(1) Types of instruments available for position fixing (compass, RADAR, Global Positioning System [GPS], echo sounder) (2) Condition of navigational environment (navigation area, available objects for positioning) (3) Visibility (4) Disturbing elements (wind and tide, electromagnetic wave propagation conditions)
4. Maneuvering (Definition)	Technique of controlling the course, speed, and position of the ship through such actions as control of rudder and the main engine.
(Main functions)	(1) Measure movement. (2) Select and choose ship operation equipment (e.g., rudder, main engine, side thruster, tugboat, anchor, and mooring line). (3) Determine operational power (under conditions of both single- and simultaneous multiple-device operation).
(Influencing factors)	(1) Maneuvering objectives (e.g., keeping course, controlling own-ship position, controlling speed, and docking) (2) Available control devices (rudder, main engine, side thruster, tugboat, anchor, and mooring line) (3) External forces (wind, tide, water depth, ship–ship interaction, bank effects)
5. Observing laws and regulations (Definition)	Techniques of navigating according to the International Regulations for Preventing Collisions at Sea, Maritime Traffic Safety Act, Act on Harbor Regulations, and other regulations.
(Main functions)	(1) Understand laws and regulations. (2) Reflect and implement laws and regulations in actual navigation.
(Influencing factors)	(1) Types of laws and regulations (2) Status of other traffic vessels (3) Condition in which laws and regulations are applicable (ocean area, weather and sea state, class of ship, principal dimension)
6. Communication (Definition)	Techniques of internal and external communication of intentions by using communication devices such as very high frequency (VHF) radio communication system and intraship telephone.

(Continued)

Table II.5.1 (Continued) Definitions, functions, and factors influencing the achievement of functions of the nine elemental techniques

Elemental techniques	Details
(Main functions)	(1) Select methods of communication. (2) Prepare information to communicate. (3) Select timing of communication. (4) Use language to communicate.
(Influencing factors)	(1) Communication partners (marine traffic centers, other ships, on the bridge, and onboard) (2) Communication conditions (emergency and standard communication) (3) Devices for communicating (e.g., flashing signal lights, flag signaling, and VHF)
7. Instrument operation (Definition)	Techniques of effectively using instruments providing relevant information for achieving techniques such as lookout, position fixing, and maneuvering.
(Main functions)	(1) Identify available instruments. (2) Understand methods of using instruments to obtain necessary information. (3) Identify characteristics of information provided by instruments. (4) Identify methods of using provided information.
(Influencing factors)	(1) Types of available instruments (2) Purpose of using information obtained from instruments
8. Emergency treatment (Definition)	Techniques of identifying emergencies with, for example, the main engine or steering gear, or ones arising in the environment surrounding own ship, and performing the necessary actions to respond to malfunctions and emergencies.
(Main functions)	(1) Identify location of problems. (2) Correct problems and malfunctions. (3) Complete necessary tasks related to abnormality and failure occurrences. (4) Identify and take appropriate measures for abnormal behavior in traffic vessels. (5) Identify and respond to abnormal weather and sea states.
(Influencing factors)	(1) Details of emergency (e.g., problems in the hull, cargo, engine, steering system, navigational instruments, and loading and unloading system) (2) Emergencies arising in surrounding ships (3) Abnormal weather and sea states
9. Management (Definition)	Technical management for combining the eight elemental techniques for safe navigation described in the previous sections, and team management for maximizing team member competency to improve team functions.

(Continued)

Table II.5.1 (Continued) Definitions, functions, and factors influencing the achievement of functions of the nine elemental techniques

Elemental techniques	Details
(Main functions)	Main functions of technical management and team management: 1. Understand management targets (bridge team management, technical management) 2. Understand management methods 3. Achieve required actions for techniques of management 4. Evaluate achieved function 5. Evaluate competency of members related to team management
(Influencing factors)	(1) Management targets (2) Competency of members related to team management

    7) Instrument operation: Technique of the effective use of instruments installed on the bridge

    8) Emergency treatment: Technique of identifying emergencies and handling them

    9) Management: Technique of managing techniques and team activities

The concept of arranging the techniques necessary for safe navigation on the basis of the functions achieved by the techniques is the concept of "functional approach" that has already been proposed by the IMO. Although this functional approach has since been used conceptually, the specific particulars have not been explained. The nine elemental techniques listed here constitute a system of techniques for ship handling organized by applying this concept.

    Table II.5.1 defines and summarizes the *techniques* and the corresponding *main functions* of the nine elemental techniques. In addition, *the factors affecting the achievement level of the intended functions* are listed. These techniques are described in detail in Part I of this book, "Techniques for Ship Handling", which the author recommends you read before reading this part. Elemental techniques 1 to 8, listed in the table, are related to ship navigation. Techniques 1 to 7 are required for regular navigation even when on single watch. Technique 8 is required for identifying abnormal situations (e.g., emergencies) and returning the affected elements to normal operating conditions. Technique 9 pertains to the technical management of techniques and personnel team management as well as techniques for the management of overall ship systems, cargo, material, and information. From the perspective of safe navigation, techniques and personnel (i.e., the team) are the primary management subjects.

Figure II.5.1 Organization of the nine elemental navigation techniques

One of the elemental techniques involved in management is the management of techniques, which is related to determining when and how to perform the other eight techniques. From this perspective, the management technique is closely related to the other eight techniques, whereas techniques 1 to 8 have no functional overlaps and are treated as independent techniques, as depicted in Figure II.5.1. In this figure, elemental techniques 1 to 8 are separated by double lines to indicate that they are mutually independent, but the relationships between management and the other techniques are interdependent and are therefore indicated by a single line. These nine techniques may be considered a system covering all techniques necessary for safe ship handling.

BTM, one of the subjects of this book, is presented as a discussion of the techniques for management categorized as team management in the elemental technique categorization system. The functions related to BTM and the theoretical foundation necessary for BTM are detailed in Chapters 7 and 8.

KEY FACTORS OF CHAPTER 5: NECESSARY
TECHNIQUES FOR SAFE NAVIGATION

The following techniques must be possessed in order to achieve safe navigation:

1) The nine elemental techniques must be executed completely for achieving safe navigation.

2) Each of the elemental techniques has its functions to be achieved.

3) Competency is ability to achieve the necessary techniques, the achievement level may change due to the effects of environmental factors.

Chapter 6

Factors in Achieving Safe Navigation

Various factors decide whether safe navigation conditions can be maintained. In this chapter, seafarers are treated as an independent element, and all other elements are treated as environmental factors that surround seafarers. In Section 6.1, all elements contributing to safe navigation other than seafarers as environmental elements are discussed. Finally, in Section 6.2, the actions of seafarers that are closely related to safe navigation are the target of discussion.

6.1 DIFFICULTY OF THE NAVIGATIONAL ENVIRONMENT

In the previous chapter, the nine elemental techniques necessary to achieve safe navigation are listed. The necessary techniques must be performed in the various situations that demand them. Consider position fixing, which is one of the nine elemental techniques. Position fixing can be broadly considered to comprise the actions involving estimation of the ship's position relative to geographic conditions. Position fixing is necessary technique to estimate ship position. The required precision and frequency of estimation differ depending on the navigational conditions. From this perspective, we find that there are differences in the required level of technique that must be achieved for objects. The precision of position fixing is one of the elemental factors determining the level of technique required for position fixing. For example, when navigating on open waters, a margin of error of approximately 3 miles is acceptable. However, the acceptable error is very small when entering a narrow channel, and thus, the geographic conditions in such a navigational area necessitate a position-fixing competency level with small error margins and high precision.

Navigational difficulty corresponds to the competency level required for the navigational environment. Specifically, under highly difficult conditions, the necessary functions can only be performed by seafarers with a high level of competency. Under open water conditions presenting little difficulty, there is less demand for precise position fixing, so the competency required is low. By contrast, in narrow waters, highly precise position fixing by seafarers with high competency is required.

The horizontal line in Figure II.6.1 represents the difficulty of the navigational environment, where the left end of the line is the origin, and the situation becomes increasingly difficult as one moves to the right, farther away from the origin. When the difficulty of the environment increases, more advanced techniques are needed to ensure safety. Thus, the difficulty of the environment corresponds to the competency level required to maintain safety in that environment.

In Figure II.6.1, the environment at point "a" has higher navigational difficulty than that at point "b". With an increase in the navigational difficulty, the competency required to maintain safe navigation increases. Therefore, navigational difficulty corresponds to the competency level required to achieve safe navigation under condition "a".

Figure II.6.1 Difficulty in navigational environments

Next, the factors that determine the competency required for the environment as indicated by this horizontal line are considered. When we replace competency required for the environment with navigational difficulty, the factors influencing difficulty become easier to identify. The following are factors that determine difficulty.

1) Ship maneuverability

   The turning radius, stopping distance, and other maneuvering characteristics of the ship being handled are directly connected to its difficulty of maneuvering, especially in water areas that are narrow or have congested traffic.

2) Geographic and water area conditions being navigated

   The expanse and form of the water area being navigated are factors that affect the difficulty of handling ship movement. In recent years in particular, waterway conditions, specifically curving, width, and water depth, have been a major factor affecting the difficulty of navigating large vessels.

3) Weather and sea state

   Restricted visibility due to fog, rainfall, and snowfall hinders lookout duties, which are based on vision and are bases of safe navigation; these obstacles make ship handling and ensuring safety more difficult for seafarers. In addition, ocean currents and large waves in narrow waters greatly magnify the difficulty of maintaining safe navigation.

4) Marine traffic conditions (types and volume of traffic)

   A large number of ships navigating in a limited water area increases the difficulty of lookout duties. Furthermore, complicated encounters increase collision risks and navigational difficulty.

5) Traffic regulations

   As the number and size of ships increase, traffic regulations have been imposed on previously unregulated ship traffic flow in order to ensure the safety of ships by implementing fixed rule–based navigation. Regulations regarding conditions in navigational areas contribute to navigation safety and decrease navigational difficulty.

6) Onboard maneuvering support systems

   Various nautical instruments, such as RADAR/Automatic Radar Plotting Aids (ARPA), Electronic Chart Display Information Systems (ECDIS), and Automatic Identification Systems (AIS), are installed on the ship's bridge in order to lessen the workload on seafarers and to increase the safety and decrease the difficulty of navigation.

7) Onshore navigation support systems

   Information on traffic vessels, weather in the planned areas of navigation, and fishing boats in operation within a specific area can be obtained from onshore support systems. This information is useful in estimating uncertain future conditions and reducing navigational difficulty.

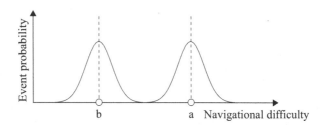

Figure II.6.2 Difficulty of navigational environment and event probability of a change in the conditions

In Figure II.6.1, the difficulty of the environment at point "a" is higher than that at point "b"; that is, the environment at point "a" demands a higher level of competency than is needed at point "b" for safe navigation. However, the navigational environment is never constant, even in the same water area. For example, the number of traffic vessels varies over time. In addition, visibility can occasionally be restricted by dense fog. Therefore, occasionally, a water area requiring the level of competency at point "b" under average environmental conditions would require a higher level of competency because of changes in the environmental conditions. Such changes in conditions can be presented using the concept of event probability. The vertical axis in Figure II.6.2 indicates the probability of changes in the conditions. When conditions at points "a" and "b" are average, the probability of changes occurring is illustrated as a probability distribution curve with centers at points "a" and "b". Of the items determining environmental difficulty, the factors that frequently alter navigational conditions must be considered. Items for which conditions might change within a single day may be considered as frequently changing items. Accordingly, factors 1), 2), 5), 6), and 7) can be considered as usually fixed conditions, whereas factors 3) and 4) are conditions that may change at any time, and so does the required competency level.

6.2 SEAFARER COMPETENCY IN SHIP HANDLING

When determining whether safe navigation is possible, both seafarer competency and environmental difficulty must be discussed as important factors. In many cases, even in environments that are difficult for inexperienced seafarers, highly competent and experienced seafarers can realize safe navigation without sensing any difficulty in the same environments; that is, ensuring navigation safety can be considered a matter of whether the seafarer's competency fulfills the competency required for the navigation environment.

The base line in Figure II.6.3 indicates the achievable competency of seafarers, where the competency increases as one moves farther to the right along the line, and the average competencies of two seafarers "a" and "b" are indicated by μ_H and μ'_H, respectively. Thus, in the figure, the competency of seafarer "b" is higher than that of seafarer "a".

The factors determining achievable ship handling competency are considered to be the following:

1) Marine license rank held by seafarers
2) Actual navigational experience on board
3) Fatigue (related to length of working hours and time elapsed standing watch)
4) Stress (related to environmental conditions for navigation and such conditions as seafarer awareness)

The vertical axis in Figure II.6.3 indicates the event probability for achievable competency of seafarers. Seafarers having the same marine qualifications and similar experience are considered to present the same competency *on average*. Accordingly, the average competencies of seafarers "a" and "b" are indicated as μ_H and μ'_H in Figure II.6.3. However, seafarers having the same seafaring license rank and similar experience do not always present the same competency. Generally, achievable competency is considered to change depending *on the individual seafarer*. Thus, the competency shown by each seafarer may be considered as being distributed within a certain range.

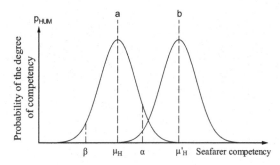

Figure II.6.3 Seafarer competency and event probability

In addition, even when focusing on a specific seafarer, his competency is changed by the influence of factors 3) and 4) in the foregoing list. Thus, changes in the achievable competency of an individual can be considered a function of the awareness of that seafarer. Thus, in Figure II.6.3, the probability of such variations is shown by the event probability of competency variations relative to the mean for that competency on the vertical axis.

6.3 CONDITIONS NECESSARY FOR SAFE NAVIGATION

In Section 6.1, navigational difficulty as determined by the environmental conditions in a specific water area was explained. Section 6.2 explained the seafarer competency to achieve the necessary functions. Next, necessary conditions for safe navigation concerning these two factors are discussed. Figure II.6.4 illustrates the relationship of these two factors, where the horizontal axis shows the competency required for the environment, and the vertical axis shows the achievable competency of seafarers. The straight 45° inclined line in the figure indicates the points at which both factors have the same value. In other words, if the state indicated by the straight line can be ensured, then *the competency required for the environment and the achievable competency by seafarers are the same*, which means that safe navigation can be realized. The region above the straight line shows where seafarers can realize higher competency than that required for the environment and consequently where safe navigation is achievable, whereas the region under the straight line indicates where seafarers cannot realize the competency required for the environment and consequently where navigation is dangerous. Hence, *this 45° line can be considered the line indicating the boundary between safe and dangerous conditions.*

Figure II.6.5 presents the variations in seafarer state and environment, described in Sections 6.1 and 6.2. The safety level of navigation may be changed due to the relationship between the competency required of seafarers and environmental conditions when changing both factors as shown.

In Figure II.6.5, when both factors are average, the resulting condition, labeled situation "A", is a state above the 45° line, which means that safe navigation is achievable. In contrast, when the environmental conditions worsen and navigational difficulty increases, the state shifts rightward to situation "B". In this situation, the competency that can be realized by

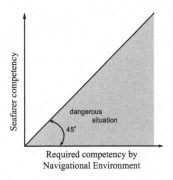

Figure II.6.4 **Necessary conditions for safe navigation according to the relationship between competency required for the navigational environment and seafarer competency**

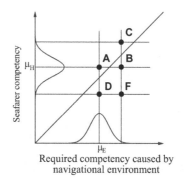

Required competency caused by
navigational environment

Figure II.6.5 Changes in ship navigation safety because of changes in environment and variations in seafarer competency

seafarers is lower than the required competency, thus indicating a change to a dangerous situation. In response, if seafarers increase their concentration and become capable of a higher degree of information processing (i.e., if seafarer competency increases), then the situation shifts to "C", which is above the 45° line, meaning that safe navigation is possible. However, if seafarers exhibit lower competency as a result of fatigue, for example, even if the environmental conditions are average, the situation shifts toward "D", which is a state under the 45° line where safe navigation may be unachievable. If seafarers show even lower competency and the environmental conditions worsen further, the situation becomes even more dangerous (situation "F"). The further the vertical distance of a state from the 45° line, the more dangerous the situation and the higher the risk of accidents.

6.4 ROLE OF THE BRIDGE TEAM IN ENSURING SAFE NAVIGATION

In the previous section, how the possibility of safe navigation can be estimated using the relationship between the navigational environment and seafarer competency was explained. Recently, it has been widely mentioned that bridge teams in which multiple seafarers are engaged in navigation must achieve the necessary function effectively in order to ensure safety. It is also pointed out that, for bridge teams to achieve functions effectively, the output of BTM and bridge resource management (BRM) must also achieve these functions effectively. Therefore, from the perspective of ensuring safety, the role of the bridge team is discussed.

The purpose of a bridge team is to ensure the safety of navigation. The reason why a bridge team realizes safe navigation can be explained on the basis of the relationship between the environmental conditions and the seafarer's competency.

In Figure II.6.6, the situation at "A" represents single watch navigation on open seas. The vertical axis represents the average competency level demonstrated for single watch navigation, and the horizontal axis represents the environmental conditions during navigation on open seas when the navigation area has few restrictions or when few ships are encountered during navigation. At "A", a competency level corresponding to the difficulty level of navigation on open seas is required. In the examples shown in the figure, seafarer competency exceeds the competency required for the environment. Thus, even a single seafarer can ensure safe navigation in these conditions.

Next, a situation in which a ship navigates a narrow water area congested by ships is considered. The competency required for this environment is higher than that in the preceding situation, so the condition shifts to the right of point "A" in the figure: a shift from point "A" to point

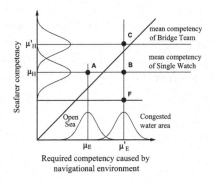

Figure II.6.6 Bridge team activities defined in terms of conditions for ensuring safe navigation

"B" because the average competency achievable by a single seafarer is μ_H. Thus, this situation falls under the 45° line, and safe navigation cannot be ensured. In such cases, the captain organizes a bridge team so that multiple seafarers can perform the functions necessary for ship navigation at a higher level than a single seafarer. At point "C", the average competency level achievable by the bridge team is μ'_H. A high level of competency can be achieved by the bridge team, so the situation shifts to point "C" above the 45° line, and safe navigation can be realized. This change in the navigation status is the basis for adopting bridge teamwork.

However, analyses of marine accidents have revealed that ineffective bridge teamwork can lead to accidents, as discussed in detail later in this book. In brief, accidents may occur when the competency level μ'_H originally expected of bridge teams is not realized (i.e., the competency level demonstrated by the bridge teams occasionally falls below point "C"). Accident analyses have revealed that in some cases, the overall competency of a bridge team at the time of the accidents was less than that expected of a single seafarer (point "F" in Figure II.6.6). The purpose of BTM is to maintain the expected level μ'_H and to maintain the necessary conditions for safe navigation (point "C"). What activities are required to maintain the expected bridge team functions? This challenge is the subject of BTM.

KEY FACTORS OF CHAPTER 6: NECESSARY CONDITIONS FOR SAFE NAVIGATION AND BRIDGE TEAM

The following conditions must be understood to achieve safe navigation:

1) The achievable competency of seafarers and the competency required by the environment are related in order to achieve safe navigation.
2) The achievable competency of seafarers must be the same as or higher than the competency required by the environment in order to ensure safe ship navigation.
3) A bridge team must be organized when safety cannot be ensured by a single seafarer.
4) The purpose of organizing a bridge team is that the competency realized by a team of seafarers is higher than that of a single seafarer.

Chapter 7

Background of Bridge Team Management

In this chapter, the concept of bridge team management (BTM) will be explained, and the background in which the necessity of BTM for safe navigation was recognized will be introduced.

7.1 DEFINITION OF BRIDGE RESOURCE MANAGEMENT AND BRIDGE TEAM MANAGEMENT (KOBAYASHI 2012)

Management activity for improving the maintenance of navigation functions on the bridge, which is the commanding place in a ship, was initially referred to as bridge resource management (BRM). However, in recent years, BTM, a similar concept, is being used increasingly. In this section, how both management approaches have received approval at international conferences is explained.

In the International Convention on Standards of Training, Certification and Watchkeeping for Seafarers (STCW) issued by the International Maritime Organization (IMO), BTM and BRM are not clearly defined.

Originally, the objective of both BRM and BTM was the maintenance and improvement of navigation functions on the bridge. However, despite organizing bridge teams in order for multiple seafarers to achieve the necessary functions for safe navigation, there have been many cases of marine accidents due to lack of team activities. Root cause analyses of these marine accidents have revealed occasional errors made by the constituent team members; that is, accidents are caused by insufficient fulfillment of the functions assigned to the individuals. The effective use of human and material resources is essential to improve the maintenance of navigation functions. However, incomplete fulfillment of functions by human resources would lead to accidents. From this perspective, the purpose of both BRM and BTM is to maintain and improve the functions of human resources on the bridge. Thus, there are clear differences between BRM and BTM.

BRM is an idea *for maintaining and improving the functions of human resources deployed on the bridge*; it emphasizes the attitude and behavior of the person in command and the management of these tasks. It considers this person to be the core of human resource management and the team leader who must learn BRM and maintain and improve functions. In many cases, this role is played by the captain.

Ensuring and improving the functions of human resources on the bridge is not possible if only the captain is responsible. The expected functions must be the achievement of each individual seafarer on the bridge. Therefore, the situation is different from being on a single watch; that is, a team of multiple seafarers must work toward the shared purpose of safe navigation. BTM ensures that *each constituent team member makes an effort to maintain and improve functions* when performing teamwork. Thus, BTM can be expressed as an essential function of team activity. In other words, *ensuring and improving the functions of seafarers on the bridge is not only the responsibility of the captain but involves the participation of all constituent team members*. One constituent team member acts as the overall leader, and in many cases, this role is played by the captain. The role of BRM is appropriate for the captain because his or her objective is the maintenance

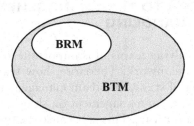

Figure II.7.1 Interrelationship of BTM and BRM

and use of human resources deployed on the bridge. Constituent team members must identify and perform the functions required to improve team functions; that is, they must know the BTM concept.

All constituent team members, including the team leader (captain), are subject to BTM. Consequently, BRM activities performed by the captain must achieve the functions required of a team leader as well as those of a team member. Thus, BRM can be considered an element of BTM.

Figure II.7.1 represents the interrelationship of BTM and BRM in the form of a Venn diagram, a tool often used in set theory. As shown in the figure, BRM is a specialized management function for team leaders and is a part of BTM.

Accordingly, *BRM training is not possible without BTM training.* The team leaders must use their team members to maintain the navigation functions through teamwork. Thus, the assigned functions must be performed by each team member in order for the team to remain functionally active; in other words, for BTM. Since the team leader is also a team member, the team leader must also employ BTM. In other words, the team leader must be proficient in both BRM and BTM.

7.2 INTRODUCTION TO TEAM MANAGEMENT IN AIRCRAFT HANDLING

In all workplaces involving teams, it is important to eliminate accidents caused by a lack of teamwork. Therefore, how team members behave has become a subject of study, and team management a topic of discussion. The necessity of team management on ships is not often discussed in regular navigation training. This is because the functions required for safe navigation require a relatively limited tasks, and if a limited number of constituent members are consistently fulfilling the assigned functions, whether the individual team members are achieving the functions for safe navigation becomes irrelevant. By contrast, the necessity of team management does arise in difficult navigation conditions, where multiple tasks may need to be completed in a short period of time. Not performing the necessary tasks in a timely manner in such instances can result in accidents. Thus, all constituent team members must fully perform their required functions. Thus, how teams fulfill their functions is discussed here.

The *adequate performance of necessary tasks* means that all tasks needed to address a given situation are performed. However, stating that a function must be *performed adequately* does not solely mean the sum of the items. "Adequately" here means that the performed task must *be both effective in and relevant to the given situation*. Thus, *the order of and coordination between the performed tasks must be appropriate to achieve the intended effects.*

Situations often arise in which BTM proves inadequate, and such situations often go unnoticed. In many cases where inadequate BTM is identified, inadequacies in team functions become evident only when the situations deviate from the norm and lead to accidents.

Consider these example cases. The first case is an aircraft accident that occurred at 17:06 (local time) on March 27, 1977, at Tenerife Airport in the Canary Islands of Spain. (Kobayashi, 2016) The necessity of teamwork was first raised in the aircraft field. The Canary Islands' international airport is in Las Palmas, but on the day of the accident, the airport was warned that extremists had planted a bomb in the airport premises. Consequently, many aircraft scheduled to land at Las Palmas Airport used the nearby domestic airport in Tenerife.

Tenerife Airport is a small airport with a single runway (Figure II.7.2) and has no ground control RADAR for monitoring aircraft in the airport. The accident occurred when a KLM plane taxiing for takeoff collided with a PAA plane on the runway. This major accident resulted in the death of all passengers and crew on the KLM flight and the death or injury of most passengers and crew on the PAA flight. Visibility at the time of the accident was 300 meters, which made it difficult for the KLM plane to detect the PAA plane at the end of the runway.

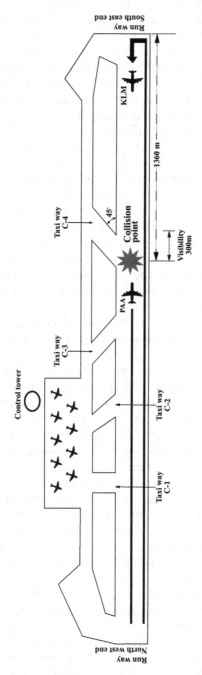

Figure II.7.2 Situation at Tenerife Airport at the time of the collision accident between the PAA and KLM aircraft

One of the factors investigated as a direct cause of the accident was the behavior of air traffic controllers. First, air traffic controllers at domestic airports are inexperienced in guiding as many aircraft as were present at the airport on the day of the accident. In addition, an international soccer game that was broadcast live on-air was in progress at the time of the accident. In the audio recordings of air traffic controllers, in addition to the conversations between controllers directing aircraft, this broadcast can be heard. That is, the air traffic controllers were performing their duties while listening to the soccer game commentary. In addition to the lack of experience, the inattentiveness of the air traffic controllers also directly contributed to the accident. Furthermore, there were problems in the communications between air traffic controllers and the aircraft they were handling.

Next, we discuss the relationship between the behavior of each of the aircraft and the accident. Both the PAA and KLM planes were in a similar situation at Tenerife Airport: both aircraft were waiting for their takeoff slot. When the KLM plane reached the end of the runway to prepare for takeoff, the PAA plane was cleared by the air traffic controllers to taxi to the takeoff point via the taxiway. Typically, which taxiway to take is dictated by the controllers, and in this case, it would have been clearly indicated as taxiway C-1 or C-2. However, at the time of the accident, the air traffic controllers simply indicated "third". This was a vague command. It is suspected that on receiving this command, the PAA plane, which had already taxied past C-1, may have been advancing toward C-4, the third taxiway.

In addition, another factor was that the PAA plane was a large-body aircraft, so it was difficult for it to proceed to the taxiway, which required a sharp turn from the runway. As shown in Figure II.7.2, taxiway C-1 requires a 90° turn, whereas taxiways C-2 and C-3 require a difficult 135° turn. It is difficult for large-body planes to turn sharply on to taxiways from narrow runways, and such movements can result in they the aircraft slipping off the taxiway. Under these circumstances, the PAA plane was heading toward C-4 taxiway to make a 45° turn.

Furthermore, a pilot, a copilot, and an engineer were present in the cockpit of the KLM plane. The pilot was an expert who had instructed for KLM's flight academy, and the copilot was a relative newcomer. The KLM plane taxied to the end of the runway and was preparing to take off by turning 180°. The accident cause analysis has indicated that the large-body plane making a 180° turn on the narrow runway may have placed a large psychological burden on the pilots. Furthermore, the pilots were under psychological stress because of changes in their company's internal rules. Right before this accident, KLM's internal rules had changed: crew overtime, previously compensated entirely by the company, was now the responsibility of the pilot if it exceeded a certain threshold. At the time of the accident, the aircraft had already been delayed by numerous hours at Tenerife Airport, so the pilot was under pressure to depart as soon as possible. In addition, because the visibility was decreasing rapidly, the pilot may have been under additional pressure to take off before the visibility was deemed too low to take off, which would have resulted in the aircraft needing to stay overnight at Tenerife Airport.

Next, the factors directly related to the accident are discussed. According to flight records, the following is what the KLM plane reported to air traffic controllers when it arrived at the end of the runway. Typically, in such circumstances, the pilot would have stated "We are ready for take-off" and waited for permission to take off. However, in this case, the pilot said "We are at takeoff". The air traffic controller is reported to have interpreted this as "We are now at takeoff position". Therefore, the air traffic controller responded "OK". On receiving this response, the KLM plane interpreted that it had received permission to take off and therefore began takeoff procedures on the runway. According to flight recorders, at this time, the engineer was advising the pilot to abort the takeoff as permission had still not been received from the controllers. However, the record shows that the pilot determined that the "OK" that had been conveyed was permission to take off. The copilot offered no further advice regarding this statement by the pilot, and the plane continued at takeoff speed. Because the pilot was an expert and the copilot was relatively inexperienced, the pilot not taking the copilot's advice seriously and the copilot not continuing to advise the pilot are considered status-based actions. Thus, the KLM plane continued with the takeoff procedures. By the time the PAA plane could be seen by the KLM cockpit crew, it was too late to abort the takeoff. Furthermore, at this stage, the KLM flight's velocity was insufficient to achieve lift; consequently, the KLM flight collided with the PAA flight.

The accident analysis revealed the following points as the direct causes of the accident:

1) Insufficient communication between air traffic controllers and the KLM plane
2) Guesswork in response to incomplete communication by air traffic controllers and the consequent incorrect decisions
3) Overly assertive behavior by the pilot of the KLM plane
4) Insufficient support actions by the copilot of the KLM plane

Thus, this accident involved various external factors. However, if the two pilots on the KLM plane had appropriately assessed the situation and fulfilled their necessary functions, such as completing the required verifications, then this accident would not have occurred. This accident is an ideal case to highlight the importance of teamwork in the cockpit. Following this accident, the airline industry reviewed teamwork in cockpits, and relevant training has since been provided for more proactive fulfillment of the necessary functions during teamwork. This concept is referred to as Cockpit Resource Management or Crew Resource Management.

7.3 INTRODUCTION TO TEAM MANAGEMENT IN SHIP HANDLING

Next, the necessary requirements for teamwork in ship handling through cause analysis of a maritime accident are considered.

The maritime accident in question for analysis is the grounding accident that occurred with the renowned passenger liner *Queen Elizabeth II*. In August 1992, the *QEII* ran aground at Vineyard Sound on the south side of Buzzard's Bay, approximately 50 kilometers east of Rhode Island on the East Coast of the United States. The following section features the accident report prepared by the U.S. National Transportation Safety Board.

At the end of the report, a nautical chart that shows the entries made by the coast guard has been included. It is recommended that you refer to this chart while reading the report.

MARINE ACCIDENT
REPORT QUEEN ELIZABETH II GROUNDING VINEYARD SOUND AUGUST 7, 1992

On the evening of August 7, 1992, RMS Queen Elizabeth II (hereinafter, QEII), outbound from Oak Bluffs to New York, ran aground at Vineyard Sound. The vessel grounded approximately 2.5 miles south of Cuttyhunk Island. Visibility was good, the sea was calm, and the tide was ebbing. The ship was carrying 1,824 passengers and 1,003 crew.

The Ship

QEII is 963′ LOA, with a beam of 105′. At the time of the accident, her bow draft was 32′ 04″ and 31′ 04″ stern. She was powered by twin-screw controllable pitch propellers and a 130,000 HP diesel electric engine. Sea speed was 32 knots, and her normal Atlantic crossing speed was 28.5 knots. QEII was the largest, fastest, and deepest ship ever to transit Vineyard Sound.

In addition to the master and the pilot, the navigation bridge of the QEII was staffed by the scheduled 8–12 bridge watch, which consisted of a first officer, a second officer, a quartermaster, and a helmsman. The helmsman was assigned to steer the rudder and took his orders directly from the pilot. There were no language barriers between the pilot and the navigation watch.

Both RADARs were in operation. The forward RADAR was available for use by the master and the pilot. The after RADAR was used by the second officer to fix the vessel's position. Two of the three echo sounders were working; the nonoperating echo sounder was the one in the wheelhouse. The two operating echo sounders, with recorders, were in the chartroom, just aft of the wheelhouse.

Chronology

Oak Bluffs was the last scheduled port of the cruise that had commenced in New York on August 3, by way of Bar Harbor, Maine; St. John, New Brunswick; and Halifax, Nova Scotia. QEII departed Halifax at 1820 on August 6, and passed east of Cape Cod, south and east of Nantucket Shoals, and south of Martha's Vineyard.

1145 August 7. QEII arrived approximately 5 miles west of Gay Head to meet the pilot boat.

1150 Pilot boarded QEII for the trip into Vineyard Sound to Oak Bluffs. Original arrangements had called for the ship to meet the pilot boat west of Buzzard's Bay Light, approximately 4 miles west of Cuttyhunk. Because the weather was favorable, and the master wanted to avoid the shoal area south of Sow and Pigs Reef, the master requested the new boarding point, and the pilot agreed.

When the pilot arrived on the bridge of QEII, the master recognized him as having piloted the vessel the previous year at Newport, R.I. The pilot was already familiar with the maneuvering characteristics of the ship. He was given a pilot card, and he inquired about the draft, vessel characteristics, and whether the engines, bow thrusters, and RADAR were working. The master did not remember whether it was at the time of the pilot's boarding or during the passage toward the anchorage that he asked the pilot where he "wanted to leave the vessel when we came out that evening". The master stated that the pilot showed him a location on the chart "roughly between Blizzard's Bay Tower Light and Brenton Reef Tower".

QEII proceeded into Vineyard Sound

1330 QEII anchored in the position previously planned off of Oak Bluffs Passengers were ferried ashore in the ship's launches.

The pilot remained on board QEII, had lunch on the Lido Deck, and then read, walked about the ship, and relaxed.

2000 Scheduled departure time from Oak Bluffs in order to make a 0700 arrival time in New York on August 8.

2050 Actual departure time from Oak Bluffs, due to a delay in boarding the returning passengers. Master had the conn and used the main engines and bow thrusters to turn the vessel to the correct heading before turning the conn over to the pilot.

The pilot testified that after the anchor was weighed, "There was a small discussion on what time we could make the pilot station and if we could run at a good speed. And the courses were fairly known without confirmation. We would primarily follow the inbound passage".

The master stated that after the ship's navigator laid out the passage plan for exiting Vineyard Sound, he approved it. The master was not aware of the pilot's plan to alter course at the NA buoy and to pass north of Brown's Ledge Shoal to get to the pilot disembarkation point.

QEII proceeded at slow speeds because of small boat and ferry traffic. She was on manual steering with both hydraulic steering engine pumps operating. Position fixings were being taken every 6 minutes by the second officer.

2115 QEII rounded West Chop, change course to 237° at buoy #26 on starboard. The pilot recalled that there was negligible gyro error and that he adjusted base course as needed to "make the buoys that I wanted". Although the navigator had laid out a trackline on the charts that had been approved by the master, the pilot indicated that "I had in my own mind my own practice of proceeding out of Vineyard Sound". He further stated, "I did not consult the navigator or the ship's charts as to what courses he was laying down versus what I was to use on the outbound passage".

2124 Speed increased to 24 knots at the master's request. To make the scheduled arrival in New York, it would be necessary to average 25 knots, and the master wanted to increase speed early in order to avoid having to use higher speeds later.

2142 Changed charts. The chart now in use did not have remarks on the shoals around Sow and Pigs Reef highlighted, but the previous chart did.

2144 QEII passed NA buoy on a heading of 235° T. With the buoy abeam to starboard, changed course to 250° T.

Note:
The approved ship's track (and, in fact, the ship) passed over a 40' depth spot near the NA buoy. Both the master and the pilot had allowed for a 40' least charted depth, and both had calculated 2 feet of squat at 25 knots and the tide to be plus 1.5 feet for the transit.

It was later reconstructed by the investigator the QEII passed over the 40' spot with 1' or less clearance under the transducer, according to the recording echo sounder.

The pilot did not inform the master or the watch officers of his intent to change course at NA buoy, nor of his further intention to change course to 270° T when south of the southwest tip of Cuttyhunk.

2148 The second officer plotted a fix and laid out a trackline of 255° T. The second officer noted that this 255° T trackline crossed over a 34' shoal area approximately 7 1/2 miles away, north of Brown's Ledge. The second officer informed the first officer of the shoal. The first officer informed the master. The master told the first officer to tell the pilot he would rather pass further south of Sow and Pigs Reef and toward the original trackline as marked on the ship's charts by the navigator.

215x Shortly before 2154 (exact time not logged), the pilot changed course to 240° T to comply with the master's request.

2154 The second officer plotted a fix and laid out a 240° T trackline. The second officer noted that this trackline passed over a 39' spot. He was not concerned because of QEII's 32' 04" draft, and said nothing to the pilot, the first officer, or the master.

The pilot said that he looked at the 240° T trackline and saw that it passed south of Brown's Ledge.

The master also looked at the chart.

Note:

According to the pilot, the predicted height of the tide during the passage out of Vineyard Sound was +1.5 feet. Both the master and the pilot testified that they considered passing over the 39' sounding with a +1.5-foot tide to not be a problem.

2158 QEII experienced severe vibrations that were felt throughout the vessel. Bridge personnel recalled two separate periods of shaking and rumblings. Bridge equipment rattled and shook in a manner similar to that experienced in rough seas. As the second vibration was ending, the master ordered the engines stopped.

The master testified that the first two reasons for the vibration that immediately came to mind were a collision with another vessel or a machinery failure; the third and only remaining possibility was a grounding.

The pilot at first suspected a mechanical failure, such as losing a propeller.

The master verified that no mechanical difficulties existed and that there were no other vessels around. Only then did the master and the pilot conclude that the vessel must have grounded.

Note:

Damage was significant. Temporary and permanent repairs to QEII cost approximately $13.2 million. In addition, the total revenue lost for the period before the vessel returned to service on October 2, 1992, was estimated at $50 million.

No injuries or deaths resulted from this accident.

Here, it is tried to extrapolate all of the elements thought to be related to the grounding accident by referencing the accident report and the nautical chart (Figure II.7.3) on which the route sailed by the QEII and the planned ship course were written.

The case is ideal for understanding the steps through which accidents occur when the duties required for teamwork and adequate functions are not fulfilled.

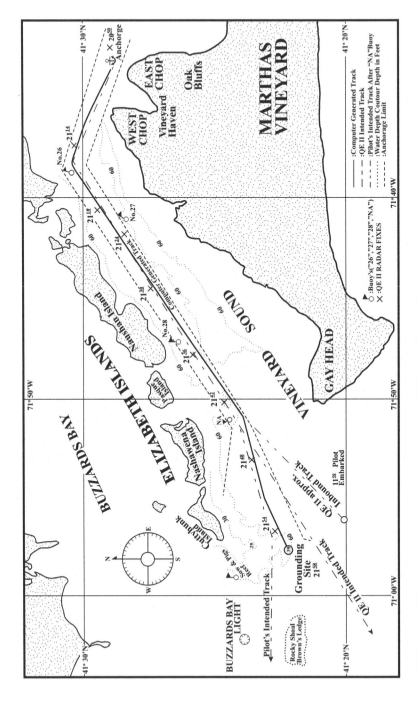

Figure II.7.3 Nautical chart showing the circumstances of the QE II grounding accident

Accident analysis

Herein, an example of the related elements extrapolated through accident analysis is shown.

- Tide level: At low tide.
- Nautical instruments: Of three echo sounders, the control room echo sounder was turned off.
- Passengers late in returning to ship (Approximately 1 hour late in heaving up anchor)

Chronological sequence of problems

1150 Inadequate verification of pilot disembarkation point.

2050 The time for heaving up anchor according to the passage planning to New York was 20:00 hours, but the passengers were late in returning to the ship, so heaving up anchor was at 20:50 hours. (This led to acceleration to 25 knots at 21:42 hours.)

2115 The pilot gave orders for the course he customarily used instead of the captain's course line. (Inadequate communication between pilot and ship is a major factor in subsequent changes in passage planning leading to accidents.)

2124 An average speed of 25 knots was required to maintain the estimated arrival time in New York. The pilot therefore accelerated to 24 knots.

2142 The nautical chart was switched, and the location of shallows near the grounding area was not marked on the new chart. (Insufficient information on risky conditions.)

2144 (Notes) The estimation of ship's sinkage was inadequate.

2148 Passing through the shallows was predicted by the 2/O, but during the interim that information was passed onto the pilot, it went from the C/O to the pilot after the 2/O informed the C/O and captain. (Late communication.)

2154 2/O noticed that a zone of water depth 39 feet (approximately 12.7 m) was again being passed through, but 2/O determined that there was no problem in the relationship of the ship's draft (bow: 9 m 85 cm) and water depth, and did not report anything to anyone. (There was no reporting of results from the assigned information collection tasks. In addition, there were independent assessments of the situation with nothing being reported to the commander.)

Although a team had been organized, there was an inadequate performance of functions that ought to have been done by the team, which resulted in the accident. By analyzing the sequence of events leading up to the accident highlighted here, what is important when a team takes action can be extrapolated. The problems in terms of team actions can be listed as follows:

1) Inadequate communication of intentions between the crew of the ship, including the master and pilot in charge of handling
2) Inadequate system for collecting information from nautical chart entries and echo sounders
3) Inadequate communication between team members and other team members

Although there may be other detailed examples (Uchino and Kobayashi 2008), the ones presented here are all highly critical items that may have been lacking in the teams' actions. Other items relevant and crucial for teamwork are discussed in the subsequent Chapter 8.

KEY FACTORS OF CHAPTER 7: BRIDGE TEAM MANAGEMENT

The following items can be learned from the background and the presented examples of BTM.

1) Management includes two ideas: BTM and BRM.
2) BRM consists of the functions that must be fulfilled by the team leader as part of BTM. Consequently, BRM is considered a part of BTM.
3) Analyses of past accidents have shown that inadequate action by team members puts the entire team at risk of an accident.
4) All team members must achieve functions in order to realize all team activities.

REFERENCES

Kobayashi H.: Advanced BTM Training beyond Revised STCW in 2010 (2012), *Proceedings of MARSIM 2012 (International Conference on Marine Simulation and Ship Maneuverability)*.

Kobayashi H.: *Techniques for Ship Handling and Team Management* (2016), Kaibundo Ltd., Japan.

Uchino A., Kobayashi H.: Function of Pilots as a Bridge Team Member from Analysis of Marine Accidents (2008), *Proceedings of 8th ACMSSR (Asian Conference on Marine Simulator and Simulation Research)*.

Chapter 8

Bridge Team Management

As explained in the previous chapter, a team is organized when there are environmental conditions in which safe navigation cannot be achieved through single watch. Consequently, a bridge team is organized in circumstances requiring a high level of competency in order to maintain safety. However, when past accidents are analyzed, there are reports in which an active team was present, but one that was functioning at a low level and thus produced results no better than those of a single watch. a bridge team occasionally lead to accidents. In this chapter, the functions required of constituent team members that are necessary for a functioning bridge team are described.

8.1 NECESSITY OF BRIDGE TEAM MANAGEMENT TRAINING

Before explaining the functions required of a bridge team, the necessity for bridge team management (BTM) training, a topic that has received much attention in recent years, is discussed. The significance of bridge resource management (BRM) and BTM were first discussed by those attending international conferences on maritime issues in 1993. A report from that year described the necessity of applying the concept of team management, which was first introduced as cockpit resource management in the airline industry, to ships. Before 1993, it was common knowledge in ship navigation that navigating was a team endeavor when the situation required it. However, there had been no specific discussion of the state of teamwork and the behavior of constituent team members. This situation arose not merely because of the absence of the concepts of BRM and BTM but because of the belief that discussion was unnecessary. This chapter will explain why a topic that did not need to be discussed before is being discussed now.

The absence of BTM training in the past may have been due to differences between the makeup of past and current bridge crews, which are as follows:

(1) **Changes in seafarer experience level**

In the past, seafarers in the deck department, especially licensed seafarers, had longer onboard experience than seafarers do today; they advanced in qualifications and rank through such experience. However, currently, seafarers advance to the top ranks with less onboard experience. Ship navigational competency is completed from knowledge through onboard experience. Seafarers with little onboard experience show a lack of competency compared with those who are more experienced, as described in detail in Part I of this book, "Techniques for Ship Handling".

Ship handling today is a composite of two types of seafarers with different onboard work experience. For them to reach a shared level of awareness and understanding, they must mutually communicate their intentions by exchanging information. When ships are underway, discrepancies in awareness and understanding between the two groups of seafarers can occasionally lead to the development of risky situations. To prevent these situations, communication is essential to ensure that awareness and understanding are shared when navigation duties are being performed. In addition, differences in awareness result in fragile work relationships among multiple seafarers. To compensate for this, training, specifically BTM training, is required.

Examples of situations arising from differences in competency

Examples of risks caused by differences in competency, focusing on differences in awareness, are introduced. The differences in awareness between an experienced captain and an inexperienced young navigator are considered.

When a very experienced captain evaluates the behavior of an assigned navigator, captain evaluates the behavior of the young officer based on captain own behavior when captain was a navigator. When captain was a navigator, captain made decisions alone on the necessary tasks and executed them. Then, captain would always voluntarily report anything that required reporting. Now, the young navigator may perform his tasks in the same ways as the captain would. This is how an experienced captain perceives the behavior of a young navigator. However, an inexperienced navigator cannot determine and execute all the tasks required for a situation. This is because even if his or her competency level was achieved through actual work experience, he or she has not been given sufficient opportunity for the experiences needed for full competency. Thus, occasionally, he or she cannot determine what items must be reported. In addition, he or she cannot determine the necessity of reporting.

A highly experienced captain, when he or she was a navigator, would have demonstrated the necessary behavior even without specific commands from the captain. However, an inexperienced young navigator relies on commands to perform the necessary behavior.

By contrast, the thought process of an inexperienced navigator would be that he or she does not have a complete grasp of what behavior is necessary, and it would therefore be appropriate for him or her to receive commands from the captain on such matters.

Ultimately, an experienced captain waits for necessary tasks to be completed without issuing commands and an inexperienced navigator does not perform tasks but waits for commands to be issued. In situations such as when encountering a ship, this relationship can lead to a dangerous situation.

An experienced captain or senior officer must thus actively issue commands to inexperienced navigators in order to realize safe navigation as well as to increase their competency level. By contrast, inexperienced navigators must learn the necessary tasks from the behavior of experienced senior officers and strive to master competency therein. Ascertaining differences in thought is a difficult matter. Consequently, both must make an effort to check their own thought processes and perceptions. If both sides fail in their efforts, then they may not be able to resolve whatever problems they are presently facing. Specific examples of such efforts can be found in situations where the behavior that BTM aims to achieve solves such problems.

(2) **Changes in seafarer thought and the appearance of diversity**

Ships today are commonly operated by seafarers of different nationalities. Thus, there exist differences in values and knowledge gaps between seafarers from different countries of origin and institutions with different education systems. However, consistent decision and action in accordance with team objectives and under the leadership of a team leader are essential for ship navigation. Behavior is required that promotes common understanding within the team, so that the inherently different senses of values and knowledge are unified and behavior based on the same decision-making criteria is maintained. BTM consists of techniques needed to achieve this goal.

Thus, new training in teamwork came to be seen as necessary to adapt to the aforementioned changes in the shipping industry.

8.2 REASONS FOR ORGANIZING A BRIDGE TEAM

Let us consider what is needed for teamwork on the bridge to proceed smoothly and efficiently. We start by first considering the reasons to organize a bridge team, and then the functions required of the team become evident. A bridge team is organized when navigational difficulty is high and the navigator must perform many tasks to ensure safe navigation. Then, necessary tasks must be performed with high precision in a short time as immediate circumstances arise continuously. By comparing the following situations as typical examples, concrete information may be found.

When a ship navigating a coastal water area with single watch enters narrow waters congested with ships, the captain gives commands for navigation by a team in order to ensure the ship's safety. This is because the oncoming water area would require highly precise, as well as highly frequent, position-fixing measurements. Furthermore, the frequency of encountering other ships in the water area would also be high, so constant observation of surrounding ship conditions and collision-avoidance maneuvers would be required. Circumstances such as this, in which a bridge team is formed, would make it impossible for a single seafarer to maintain the necessary quantity and quality of task assignments.

Organizing a bridge team in response to such a situation enables the many necessary tasks to be achieved by multiple seafarers. Thus, the number of tasks required of a single seafarer is reduced, and each individual constituent member is able to accurately complete their assigned tasks respectively. The team is able to realize safe navigation by utilizing the tasks performed by each constituent member. This is the reason for and the objective of organizing a bridge team.

8.3 SPECIAL ASPECTS AND NECESSARY FUNCTIONS OF TEAMWORK

The qualitative difference between single watch and bridge team operation is that when on single watch, all necessary tasks are performed by one seafarer. By contrast, with a bridge team, the necessary tasks are assigned to multiple seafarers. When we examine the qualitative differences between single watch and bridge team, clarifying these differences also clarifies the techniques that must be achieved by the bridge team.

With bridge teams, the necessary tasks performed by the multiple seafarers are essential ones indispensable for safe navigation. Consequently, the results of all tasks performed by the seafarers must be reflected in the handling. That is, the individually performed tasks must be integrated and then analyzed and evaluated for ship handling. The process of integration usually takes the form of *a task execution results report*. That is, information transmission as *communication is required*.

If a single seafarer is handling a ship, then all tasks are performed by the sole seafarer on duty; thus, the results are known to that person. Information transmission and communication are not needed; therefore, information transmission as communication is considered a specific activity of bridge teamwork.

Furthermore, tasks assigned to multiple seafarers must be performed smoothly as team activities. Thus, tasks cannot be done haphazardly among the team members, and each team member must grasp the circumstances of other members' actions. This type of behavior is termed *cooperative behavior*. Cooperative functions contribute to connectivity in the assigned tasks performed by the multiple seafarers and are thus essential for team activity.

Cooperative behavior and communication: these two functions must be performed by all constituent team members. In addition, for these functions to be fulfilled, other important functions are required to facilitate teamwork. These *functions are supervising the entire team, analyzing and evaluating results from tasks performed by constituent team members, and taking responsibility for deciding and guiding behavior for safe navigation*. These are required functions of the team leader, and their fulfillment is linked to the achievement of team objectives.

Next, the significance of cooperative behavior, communication, and management by the team leader is explained.

KEY FACTORS OF SECTION 8.3: NECESSARY FUNCTIONS OF TEAMWORK

In order to achieve the purpose of organizing a bridge team with plural seafarers, team members must achieve the following functions:

1) The team leader must achieve the function of motivating the team so that it can achieve its purpose.
2) Team members must communicate effectively.
3) Team members must maintain cooperation in order to maintain smooth team activity.
4) Team members must properly perform the work assigned to them.
5) The team leader must achieve items 2 to 4 as a member of the team. Team members must also follow the intentions of the team leader.

8.4 COMMUNICATION

In general, communication is understood to be interpersonal conversation. However, on the bridge this is done through commands and reporting. General conversation on the bridge is not part of the communication needed for teamwork. Here, the significance of the communication that is considered essential when organizing a team is considered. The function of communication is the mutual exchange of information held by constituent team members, and is thought to accomplish the following:

1) To integrate the results of tasks performed by each team member (constituent team members)
2) To provide team members with the same information and the same sense of purpose
3) To enable breaking of the human error chain

As described earlier, the results of duties performed by multiple seafarers must be reflected in ship handling. Information obtained as a result of tasks completed by multiple seafarers must be collected as per the person in charge of ship handling. That is, functions are required to integrate information for the operational leader of the ship. This may be considered the primary significance of communication.

Next, it gives the team members a shared sense of purpose. This is achieved by sharing the information between team members. By sharing information, all team members understand the situation the team faces, and each one can consider responses to the situation being faced. Everyone is informed of information obtained by individual team members so that it may be shared between them. Team members, including the team leader, are notified of obtained information and ascertained situations. In addition, the team leader must inform all team members of decisions he or she has made as leader. This is because all team members must behave in a manner reflecting the team leader's intentions. Thus, the team maintains functions for achieving objectives as an integrated unit.

The author has had the opportunity onboard an ocean-going vessel of examining situations in which a team was active on the bridge. The ship on which the author was onboard had a Japanese captain and first officer, but the third officer and three crewmen assigned to the bridge were Filipino. While passing through Singapore Straits, the ship entered into a risky situation with the other ships in the area. At that time, when a Filipino crewman on lookout first detected one of the risky ships, he went to the captain and notified him by pointing in its direction. The first officer monitoring the forward direction could not know this information. In addition, the third officer monitoring the RADAR was not notified. This is a typical example in which there was no shared awareness of the situation among all team members. As I heard later, Filipinos are often shy and do not raise their voices when reporting. Although I understand the situation, it was decided to point out the need for improvement.

The third point indicates that communication can break the human error chain. Usually, when a person makes an error, they do not notice it by themselves. Serious accidents are often caused by human error. Occasionally, simple human error is the direct cause, but in many cases, an initial simple human error leads to a subsequent error in judgment and then another human error, eventually creating the conditions for a serious accident. Accidents may be avoided through early detection of human error and restoring normal conditions. However, how might one make an early detection of a human error occurring? As described, usually, when a person makes an error, they themselves do not notice it. If the person who made an error does not notice it, then it is important for other people to detect the error. For the errors of others to be known, it is important to create opportunities for people to know the action and information achieved by the person who made the error. This is accomplished by making one's behavior and awareness known to others orally. Thus, the people making up a team can point out errors through constant attention to these details and listening to information carefully provided among team members. Reporting by team members should not be thought of as something done only for the team leader. Team members must pay attention to what is being said among themselves. If someone other than the person who said something notices a mistake in what was said, errors might be corrected by pointing it out. Thus, subsequent errors in judgment and human errors due to a simple human error are eliminated, which thereby eliminates the conditions for a serious accident to occur. In other words, clear communication can be considered a way of breaking the human error chain.

Human errors may arise from any team members, including team leaders. Here, examples seen when examining characteristics of seafarer competency using a ship handling simulator are introduced.

Example 1: Case on the bridge of ship westbound in the Singapore Straits

When a ship is navigating westbound in the Singapore Straits and heading north of Buffalo Rock, and encounters a ship on an eastbound deep water route heading to Buffalo Rock as in Figure II.8.1, typical conversations on the bridge are frequently observed. The seafarer on lookout mistakes the eastbound ship for a crossing risky ship and notifies the captain of a collision risk. When this happens, the captain or another team member, as a result of being notified of the high possibility that the oncoming vessel is in an eastbound lane, subsequently navigates with their attention on the target. By verifying the position of the other ship, it becomes possible to verify that the target is eastbound. However, if this correction is not made by other seafarers, then the team makes wasteful contact by VHF wireless telephone to confirm

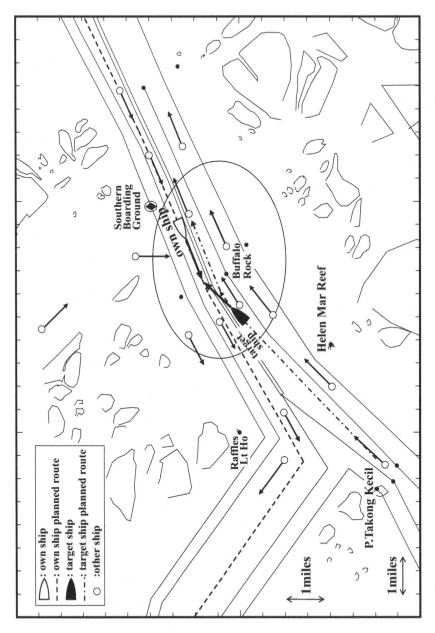

Figure II.8.1 Traffic flow characteristics in the vicinity of Buffalo Rock in the Singapore Straits (when westbound)

the other ship's intention or unnecessarily alters heading to avoid a collision. This would then be a case of error caused by a seafarer who did not fully understand the circumstances of a ship's course in the Singapore Straits, which broadly corresponds to human error.

Example 2: Case on the bridge of a ship eastbound in the Singapore Straits

A ship is on a deep water route eastbound in Singapore Straits. When it proceeds into a route crossing area toward the Jong Fairway shown in Figure II.8.2, the following conversations take place concerning a ship on the starboard bow going eastbound in the shallow waters lane. The ship being handled has already been overtaken and passed by the starboard eastbound ship, which is faster. At the point of being overtaken, both ships communicate via VHF wireless telephone and mutually verify their navigation plans. During those communications, it becomes known that the target ship plans to pass through the Jong Fairway and then Singapore Harbor. At the time of passing Buffalo Rock, the captain mistakenly guesses that the target ship is continuing eastbound, and notifies all team members of his or her intention to accelerate to full speed from half speed. When this happens, one of the team members, the second officer, notifies the captain that the target ship going eastbound in starboard shallow waters lane plans to pass through the Jong Fairway and harbor in Singapore, and acceleration is stopped. This case indicates that team leaders can also make errors, and the importance of team members always verifying the intentions of the captain.

From the two examples shown here, communication demonstrates the important function of making sure human errors do not lead to accidents. The importance of communication and methods of communication should be fully understood. This point is discussed in Section 8.9 of this chapter, "Methods of Communication".

KEY FACTORS OF SECTION 8.4: SIGNIFICANCE OF COMMUNICATION

The following functions are possible through proper communication:

1) Sharing of results by each team member, summarizing of obtained information, and determination of a teamwork plan.
2) Sharing of a sense of purpose by team members.
3) Detection of human error by a team member can break the human error chain.

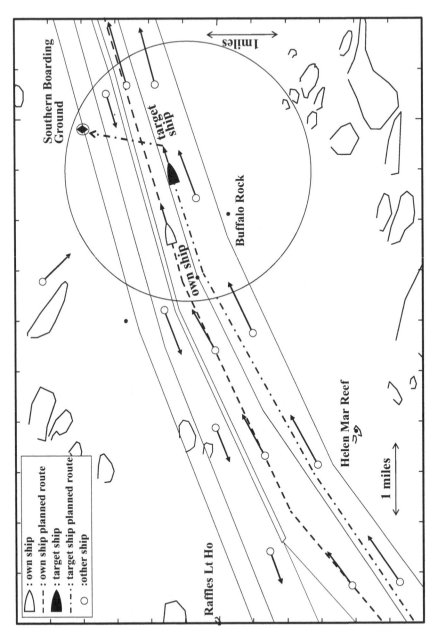

Figure II.8.2 Traffic flow characteristics of an eastbound in the vicinity of Buffalo Rock in the Singapore Straits (when navigating eastbound)

8.5 COOPERATION

Next, the significance of the cooperation necessary between team members for team activities is considered. The following is a summary of the functions achieved through cooperative activity and the necessary behavior to achieve it.

1) Realization of cooperative behavior between team members
2) Backup for necessary tasks that other team members cannot or would not accomplish
3) Verification of the behavior of other team members. Detection of human errors

Here, the cooperative functions required for team activities is explained, including examples.

(1) **Cooperative behavior between team members**

First, cooperative behavior between team members is explained. With teamwork, the tasks that must be performed by the team are assigned to multiple members. Necessary tasks are not completed by a single member, but are usually completed through a combination of the tasks assigned to the multiple seafarers. Therefore, it is important for individual seafarers to verify the results of other team members' tasks. And individual tasks are performed based on the results of these other tasks. This is the manner in which teams must work in order to complete necessary tasks.

For example, what cooperative behavior is required when another ship is detected visually and reported to the captain? Typically, if a seafarer is on single watch and visually detects another vessel, he or she collects more detailed information using RADAR/ Automatic Radar Plotting Aids (ARPA). This action is to obtain information from RADAR/ARPA that cannot be obtained visually. Information collected from RADAR/ARPA is needed to more accurately recognize the status of the detected ship. Consequently, the cooperative behavior necessary when other ships are detected visually and reported to the captain is for the seafarer on duty for monitoring RADAR/ARPA to provide detailed information acquired by RADAR/ARPA. Thus, to realize this procedure, reports to the captain by other seafarers must be verified, other ships immediately recognized by RADAR, and necessary information acquired and reported to the captain. This is considered cooperative behavior between team members.

Let us see what happens when this cooperative behavior is not achieved. After another ship is detected visually and reported, a seafarer responsible for monitoring RADAR/ARPA may catch the same ship on RADAR later. Then, the seafarer on RADAR/ARPA duty notifies the captain of the presence of the ship the other seafarer found. What does the captain determine now? First, when the captain receives the report on the visual detection of the other ship, he or she waits for a detailed report. However, if the seafarer monitoring the RADAR/ARPA does not provide information, time passes without an accurate grasp of the situation. Then, if a notification is made a short time later by the seafarer monitoring RADAR/ARPA of the presence of the other ship, this may be interpreted as the appearance of a different ship. Incomplete cooperative action induces a lack of smooth teamwork to perform assigned tasks, and may lead to inefficient work and even problems.

The fact observed in my onboard inspection at Singapore Strait, that information was not provided by a seafarer monitoring the RADAR/ARPA after initial visual detection due to the shyness of a Filipino crewman, corresponds to this example of these communication points with lack of cooperation.

(2) **Backup for necessary tasks**

Next, the function of backup for necessary tasks that other team members cannot do is explained. Here, backup means performing a necessary task in place of another team member who is unable to do it and fulfilling necessary tasks as a team. In regular teamwork on a bridge, various duties are assigned to individual team members. Depending on conditions at sea, tasks may be concentrated on a specific seafarer. Since concentrated tasks consist of individually required necessary tasks, all tasks must be done. However, occasionally, an assigned person cannot perform all of the tasks given to him or her. Under such circumstances, other members must perform tasks that cannot be done by the assigned person instead, in addition to their own assigned tasks. This behavior enables the completion of necessary tasks by the entire team. Successful teamwork requires the achievement of all necessary functions by the team. Necessary tasks may change constantly and must be performed by one of the team members, even if they exceed their initially assigned duties. This constant need to perform a task at any time is a requirement of teamwork and is important for cooperative action.

For example, let us assume that position fixing at fixed time intervals and external communication via very high frequency (VHF) radio communication system are assigned to the second officer. If the second officer is maintaining communication with a ship presenting risks or the marine traffic center, and a need arises for communication via VHF wireless telephone for a long period of time, then the second officer will not be able to perform the initially assigned task of position fixing. Under such conditions, safe navigation requires another member to perform position fixing instead of the second officer. The backup provided by performing such necessary tasks instead of the assigned member, when they might otherwise be omitted, may be considered an important function of cooperative activity.

(3) **Verification of team member behavior**

Next, the verification of one's behavior by other team members, which is one of the cooperative functions, is explained.

With teamwork, multiple members perform the assigned tasks. By verifying the behavior of other team members, it is possible to recognize when another team member has mistaken a situation or has made an error. As previously explained in the communication points, it is very difficult for a person who has made an error or misunderstanding to see it themselves. Even if a person orally conveys the gist of their thoughts and behavior as indicated in the communication items previously described, as long as they are not verified by other members, it is not possible to stop the human error chain. Accordingly, one of the essential points of cooperative behavior is the verification of other team members' behavior.

The verification of other team members' behavior is a necessary function in order to be aware of cooperative tasks that are not being performed by other members and to provide backup for them. When team members are focused only on the team leader, it may not be possible to be aware of other team members' behavior and statements. Thus, each team member must know the current status of the team.

Accordingly, it may be understood that communication and cooperation are two functions that are important for constituent team members to perform team activities.

These two functions are techniques that are important when performing team management as a constituent team member, and practicing these techniques is the objective of team management training.

**KEY FACTORS OF SECTION 8.5:
SIGNIFICANCE OF COOPERATION**

The following functions can be achieved through cooperation:

1) Maintenance of relationship in work performed by each team member and smooth teamwork
2) Substituting or supplementary work for other team members
3) Monitoring of actions by other team members and detection of human error

8.6 NECESSARY FUNCTIONS OF A TEAM LEADER

In the previous section, the important functions that team members and team leaders must accomplish for team activities are explained. The importance of these functions is not evident when on single watch. In other words, they are not necessary under single watch conditions. However, they are extremely important when engaged in team activities. The functions indicated in the previous section were ones that must be performed by each constituent team member. However, perfect teamwork cannot be realized only by these functions. There are even more important functions, and these are management of the entire team, analysis and evaluation of results from tasks performed by constituent team members, and responsibility for determining and guiding behavior for safe navigation. These are the functions required of team leaders, and the achievement of these functions is essential for achieving team objectives. Next, let us summarize the necessary functions of team leaders.

In this section, the necessary functions of team leaders charged with the responsibility of managing teams and fully achieving the tasks required of teams are described.

(1) **Team leaders must clearly notify all constituent team members of the objectives to be achieved by the team, and indicate specifically the behavior necessary to achieve those objectives.**

Constituent team members fulfill their individual tasks according to the intentions of the team leader. Therefore, the team leader must accurately and clearly communicate leader's own thoughts to constituent team members. The team leader must accurately understand what must be achieved by the team, prioritize targets necessary to achieve objectives, and propose strategies for achieving individual targets. All targets that must be individually achieved are deemed matters necessary to achieve objectives. Consequently, these targets are decided on by drafting plans for achieving objectives; that is, strategies must be considered. When safe navigation is the objective, strategies must match proposed passage planning, the individual matters required by passage planning must be broken down into individual targets, and solutions to those matters must be planned out as tactics. Leaders must keep this concept in mind and specifically indicate the behavior necessary to achieve objectives.

(2) **Team leaders must evaluate the competencies of constituent team members and clearly give orders regarding assigned duties, and the details of those duties, to all constituent members.**

Team leaders must make it their objective for the team to constantly achieve maximum results. Thus, the maximum competencies of constituent team members must be applied in order to achieve team objectives. Usually, the competencies of constituent team members are not uniform; there are differences among them. Furthermore, the duties that must be fulfilled are many and varied in their level of difficulty. Team leaders must evaluate the *individual competencies* of constituent team members, weigh the *difficulty of duties,* and then assign the duties to individual members based on the *importance of duties* to be achieved to a certain degree for safe navigation.

Inexperienced young navigators are occasionally assigned important tasks for training purposes. In such cases, situations occasionally arise in which navigational difficulty increases, and the team must achieve a high level of competency. At those times, considering safety, it is inappropriate to assign young navigators having limited competency to such roles. Teams must always achieve the maximum competency needed. From this perspective, even if the purpose is to give a young navigator a chance for training, the originally intended task assignments must be clear. In addition, if a young navigator is in training, and a situation arises in which navigational difficulty increases and the team must achieve a high level of competency, there must be a system to ensure that the team will immediately revert to normal assignments. Thus, team leaders must always keep assessing the situation and change assignments when necessary.

(3) **The team leader must constantly monitor the behavior of constituent team members, motivate teamwork, and always maintain optimum activity.**

Team leaders must constantly motivate the team and maintain optimum activity. The behavioral status of constituent team members may be assessed according to the *degree of achieving assigned tasks* and the *frequency of reports,* so team leaders should constantly monitor the behavior of constituent team members. One function of team leaders is to give advice and commands in order to promote the achievement of necessary tasks and thorough reporting whenever there are inadequacies in these points.

(4) **The team leader is also a member of the team, and so he or she is required to have the competencies required of team members described in previous sections and listed in the following text.**

The competencies required of team members shown here are the competencies that must be similarly satisfied by team leaders as team members.

1) Team members must communicate effectively for all constituent team members to share information.

2) Team members must always demonstrate cooperative behavior in order for the behavior of individual persons within the team to smoothly facilitate teamwork.

KEY FACTORS OF SECTION 8.6:
FUNCTIONS OF TEAM LEADER

1) The team leader must *clearly notify all team members of the purpose* to be achieved by the team and present the *specific actions necessary* to achieve the purpose.

2) The team leader must evaluate the competency of team members and *provide clear instructions of the duties and specifics of work* to all constituent members.

3) The team leader constantly *monitors the actions of team members* and *motivates* the team to *maintain constant optimum activity.*

4) The team leader is also a member of the team, and thus must provide good *communication* in order that team members can understand the situation the team faces, and achieve *cooperation* in order to smoothly proceed with teamwork.

8.7 EXAMPLES OF TEAM ACTIVITY IMPLEMENTATION

The state of training for the development of BTM competency is being continuously examined and researched at the Ship Maneuvering Simulator Center of the Tokyo University of Marine Science and Technology. Figure II.8.3 shows one example of bridge assignments during BTM training.

This figure shows the assignment of roles to achieve functions in the form of a three-person team consisting of two navigators under a captain. Therefore, the quartermaster is only responsible for steering the rudder and is not performing lookout duties. This is thought of as a case in which three seafarers, including a captain, are required to achieve the necessary functions of a bridge team.

The role assignments according to the diagram are a captain responsible for overall management as leader and lookout, a second officer responsible for determining mainly position fixing and lookout through RADAR, and a third officer responsible for visual lookout and communication outside the ship being handled. The sum of duties assigned to the captain, second officer, and third officer are the duties performed by a single navigator when on single watch, and enable the basic functions necessary for safe navigation.

By contrast, when a team is organized to perform duties, communication and cooperation functions are newly added. These are considered specific functions of teamwork. The functions shown in the two frames in the middle of Figure II.8.3 are important for BTM. There are no team activities without communication and cooperation. As previously mentioned, even if each of the constituent team members adequately achieves their assigned functions, this alone will not ensure safe navigation. The performance of all necessary functions starts from the integration of each of the assigned tasks. Thus, there are no team activities without communication and cooperation functions.

This diagram makes it clear that communication allows information held by individuals to be shared by all team members, and cooperation facilitates integrated behavior and backup of any lacks.

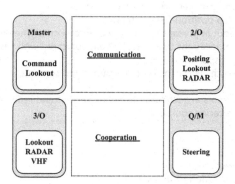

Figure II.8.3 Example of functions in bridge team activities

8.8 CAPTAIN'S BRIEFING

When necessary tasks are executed by a team, first, the team leader must explain to team members the key points and precautions of achieving their duties. Through the team leader's explanation, the team becomes integrated, and the action plan of the leader's objectives becomes known throughout the team. What is important for teamwork is that necessary functions are achieved through multiple workers, and the achieved functions contribute to the effective achievement of objectives. To make this possible, the objectives of team activity and the functions necessary to achieve those objectives must be clarified, and all constituent team members must understand the methods to achieve those functions. This process takes the form of a preliminary explanation by the team leader to team members. In BTM, this takes the form of a "Captain's Briefing", a necessary process for the team to achieve its objectives.

Table II.8.1 shows the main points of captain's briefings to be verified for team activity on the bridge.

The first point is an explanation of objectives and the strategies used to achieve those objectives. The primary objective is the achievement of safe navigation. To achieve this objective, there must be an explanation of a plan for predicting the various environmental conditions that will be encountered and basic responses to the situation.

The second point is the assignment of roles that must be fulfilled by individual team members. Oncoming sailing conditions must be forecasted, the details of necessary duties predicted, and then tasks clearly assigned to

Table II.8.1 Examples of items to be confirmed at captain's briefing

Examples of Captain's Briefing Requirements
1) Explanations of objectives and strategies for achieving objectives
2) Assignments of roles that each team member must achieve
3) Explanations of essential points in order for assigned roles to be achieved
   The following are examples:
   • Summaries explaining predicted traffic conditions
   • Important points in performing lookout duties
   • Requested ship position-fixing intervals
   • Warnings about important traffic flow characteristics when navigating
   • Verification of locations that must be communicated with vessel traffic centers
   • Verification of special local rules
   • Responses to changes in predicted weather and sea state
4) Passage planning
   The following points in particular should be verified and reflected in planning:
   • Objects for altering course
   • Heading targets
   • No-go areas
   • Safe distance
   • Under-keel clearance
   • Plan for deceleration of ship speed
   • Other important items for ensuring safe navigation

team members for all duties without any omissions. This is a basic requirement of team activity.

Third, the essential points for achieving tasks are verified. For example, lookout duties require the indication of crossing ships, anchoring vessels, and fishing vessels based on the characteristics of marine traffic conditions in a navigation area, and the confirmation of necessary lookouts and reports. In addition, position-fixing intervals must be verified by taking into account the navigable widths and the effects of tide and wind pressure.

Fourth, a navigation plan is proposed by which performing duties are decided. This corresponds to passage planning in BTM. Passage planning is governed by Safety Management System (SMS) manuals proposed by shipping companies and involves the establishing of course lines, no-go areas, heading objects, and targets for altering course. At this stage, the information collected must be used to realize safe navigation. For example, the local rules peculiar to a region must be verified by using external resources such as VTIS, sailing directions, and information charts. In addition, responses to emergencies must be verified beforehand.

8.9 METHODS OF COMMUNICATION

An important element of realizing safe navigation is the prevention of collisions and groundings, which are serious marine accidents. The important techniques for responding to these types of accidents are the essential elemental techniques of lookout duties and position fixing.

In this section, the important points for reporting are summarized:

(1) Events to be reported
(2) Information to be included in reports
(3) Reporting timing
(4) Order of reporting information
(5) Reporting frequency

The points highlighted here are the basic requirements that must always be kept in mind by seafarers when reporting. According to the BTM/BRM training experiences, there are actually cases in which many seafarers show inadequacies in reporting methods observed during team activities, even if they are experts. Young seafarers as well must pay close attention to these points. Here, I explain the points in detail.

(1) **Events to be reported**

Events to be reported means that selection should be done on the *issues to be reported* during ship navigation. Occasionally, assigned personnel do not recognize the need to report the results of assigned tasks during team activities, such as information obtained from a lookout, ship position obtained from position fixing, or the effects of outside disturbances. Cases have also often been observed in which another ship is detected by a lookout but is not reported, or a lateral deviation occurs from the planned route and its position has already been measured as the boundary of the fairway by position fixing, but the report is only deviation from planed line.

(2) **Information to be included in reports**

Information obtained from performing tasks is usually not singular. All information necessary for ship navigation must be obtained and then reported without omission. Therefore, necessary information must be recognized, tasks to obtain it performed, and then appropriate selections made from all the information obtained. Furthermore, when necessary, obtained information must be analyzed, made into information directly useful for ship maneuvering, and reported.

(3) **Reporting timing**

Generally, the appropriate timing to report information is immediately before it is needed by seafarers maneuvering the ship. If information is obtained later than it is needed, it is useless, but it is also inappropriate for it to be obtained too early. Although it depends on the event reported, events that are reported early should not be reported only once, but are preferably reported again and repeatedly in the same manner. In this case, the first report should be to notify the ship operator of warning information in advance while there is still time to spare.

(4) **Order of reporting information**

The order of the multiple pieces of information contained in a report is crucial for those people who receive that information. If an inappropriate order is applied, then the information will not serve as effective communication and thus may not be effectively used or may cause confusion. Reporting information should have a logical relationship with causality for easy understanding. In addition, the first report includes essential points when there are multiple pieces of information with varying levels of importance.

(5) **Reporting frequency**

If there are items contained in information that changes over time, then the frequency of reporting is important. Important information from lookout duties and position fixing, in particular, changes over time. It is necessary to note ways of reporting that include comparisons with previous reports and the degree of change in order to clarify how much has changed.

The respective methods of communication for achieving lookout techniques and position-fixing techniques are explained as the subsequent issues to the following "Key factors of Section 8.9".

KEY FACTORS OF SECTION 8.9: METHODS OF COMMUNICATION

The following points should be considered when communicating with team members:

1) Ensure clear awareness of the purpose of communication
2) Classify information and items to be communicated
3) Select the timing for communication
4) Decide on the order of information to be communicated
5) Decide on the frequency of communication

Communication for lookout duties

The functions to achieve lookout technique are described as follows in Part I of the book.

1) Recognition of present situation

Kinds of target ships, types and movements (i.e., position, course, and speed) of ships encountered

2) Prediction of future situation

Movement of targets (future position, course, and speed), changes in movement of targets, and estimation of risks to the ship being handled (hereafter, "the ship") (specifically, closest point of approach [CPA], time to closest point of approach [TCPA], and bow crossing range [BCR; i.e., range from bow or stern])

Information on present situation and estimation of future risks is important. To maneuver to avoid collisions, the target ships must be detected early, and predictions for the future situation must be made accurately.

The first information needed for the present situation is the *position of target ships* as seen from the ship. The next thing needed is information on the *movement of target ships*, which is information on target *course and speed*.

Then, to predict the future situation, information is needed on the CPA (also called distance at the closest point of approach) and the TCPA and BCR for the situation of risk presented by target ships. Furthermore, if the ship engages in collision-avoidance maneuvers, the information needed to make a plan for maneuvering must be provided.

Reports concerning lookout duties always require information on the present situation and prediction of the future situation. If ships are considered the subjects in lookout information, then information on traffic vessels may be broadly divided into information on ships' general movements and information on ships at risk of collision. In particular, if reporting timing is taken into consideration, information on targets presenting a risk of collision may be divided into *when detecting ships* presenting a risk of collision, *during continuous monitoring, immediately before start of collision-avoiding action, during collision-avoiding action,* and *at the end of collision-avoiding action.* Let us consider reporting information in this order.

(1) **When detecting ships presenting risk of collision**

Current condition information consists of the *positions of target ships,* and *target ship movement information* consists of information on the *course and speed* of target ships and *crossing angle* between the ship and the target.

Information for predicting the future situation consists of CPA and TCPA. It further includes BCR, which is a value important to navigation that indicates whether the bow or the stern of the ship is being passed. CPA values are an index of safety indicating at what distance both ships are passing. However, although it is important for collision-avoidance maneuvers, it does not suggest actions for collision avoidance. When BCR means that a target ship is passing the bow of the ship (usually, BCR is indicated as a plus), the CPA increases if the ship heading is changed in the direction of the target ship, leading to a course that would pass the stern of the target ship. If BCR is a minus, that is, if it means that the target ship will pass the stern of the ship, the CPA will increase if the ship heading is changed in the opposite direction of the target ship, leading to a course that will pass the bow of the target ship.

When a ship presenting a risk of collision is first detected, and collision-avoiding action is taken based on the present situation and future predictions, it is appropriate to report BCR indicating the effects of that action. In addition, even if the target ship is the give-way vessel, the behavior of the target ship can be predicted by reporting BCR.

(2) **During continuous monitoring**

The purpose of reporting during continuous monitoring is to inform of changes in situation. If there is a continuing risk of collision, this is reported together with the distance to target, TCPA changes, and BCR as the present situation of the target. When changed distance and TCPA are reported, additionally reporting differences since previous reporting supports the actions of the ship operator.

The frequency of reporting during continuous monitoring performed from the time a target ship is initially detected to the start of action is once or more, and if there is time to spare, then reporting must be done multiple times. In addition, reporting on the situation of surrounding ships also enables consideration of new risks caused by collision-avoiding action for the current target.

(3) **Immediately before starting collision-avoiding action**

Reporting immediately before starting collision-avoiding action is needed for the seafarers deciding and executing it to verify that the decision to take action is valid. First, the target ship's position and TCPA are reported as changing information. Then, BCR, CPA, and the target ship's course speed and crossing conditions are reported as information that is not changing. In addition, depending on the situation, it is also useful to report BCR for a new course by suggesting a course to avoid.

(4) **During collision-avoiding action**

The primary purpose of reporting during collision-avoiding action is to verify whether action is changing under effective conditions. Occasionally, depending on the behavior of the target ship, collision-avoiding action would not be effective, which is another point that requires attention and reporting. The behavior of other ships that are not targets must also be monitored from the same perspective and reported as needed.

(5) **At the end of collision-avoiding action**

At times when collision-avoiding action functions are effective and the risk of collision has been eliminated, CPA must be stated, in addition to reporting that the target ship is now moving away from the ship.

KEY FACTORS OF SECTION 8.9: METHODS OF COMMUNICATION FOR LOOKOUT DUTIES

In lookout duties for collision avoidance, reporting items change at each of the following steps, so reporting must be done accordingly:

1) When detecting ships with risk of collision
2) During continuous monitoring of ships with risk of collision
3) Right before starting collision-avoiding action
4) During collision-avoiding action
5) At the end of collision-avoiding action

Communication for position-fixing duties

The functions to achieve the position-fixing techniques are described as follows in Part I of this book.

1) Selection of methods to collect information for position fixing (selection of measuring instruments and selection of targets for position fixing)
2) Position fixing (realization of required precision and frequency)
3) Estimation of movements of the ship being handled (hereafter, "the ship") (estimation of direction of movement, speed of movement, rate of turn, and effects of wind and tide)

The methods for position fixing mentioned in 1) are required so that seafarers receiving reports of information can make estimates on the accuracy of the information. Thus, successive reporting is recommended.

Let us consider the following reported information.

(1) **When navigating on a straight planned course line**
   Information on present status in position fixing includes deviations from planned course lines, leeway, and remaining distance to both sides of the fairway and no-go areas. Reporting must always consider the effect on ship position of external forces such as wind and tide.

(2) **When navigating on a planned course line including waypoints**
   When navigating a straight course with planned waypoints ahead, the direction and distance of the next waypoint must be reported together with the information reported when navigating on a straight planned course line in (1). Furthermore, there are cases where it is appropriate to make course changes by changing the planned waypoint based on the present ship position obtained from measurements. At such times, in addition to reporting a new course, occasionally

it is recommended to advance straight ahead on the present course without making changes as a result of delaying or setting ahead the time to change course. In other words, a new course line is taken without passing through the planned waypoint at all. When such measures are employed, full consideration must be given to the depth of nearby waters and avoiding entry into risky areas. Since the originally planned course took into consideration all requirements, such as risky areas, changes should be made judiciously. Sticking to a planned course without necessary thought would lead to drawbacks such as frequent course corrections and speed reduction. This point should be considered in reports related to measuring ship position to suggest a navigation plan.

(3) **When advancing toward a destination with a predetermined arrival time**

In addition to (1) and (2), if the destination ahead has a predetermined estimated time to arrival (ETA), the following information must be reported. There must be a report of the distance and direction to the destination, the remaining time until the ETA, and the speed to arrive at the expected time (expressed as required speed).

This information is reported as position measurement needed for handling the ship. The reporting timing and reporting frequency are determined by the time and frequency needed to measure ship position. The frequency of basic ship position measurements is determined by the width of the navigable area and the effects of wind and tide. If the navigable area is narrow, and strong external disturbances are predicted, then methods enabling position measurement of the ship within short periods of time using heading targets and parallel indexing must be added. This would make it possible to report ship position at the proper time.

KEY FACTORS OF SECTION 8.9: METHOD OF COMMUNICATION FOR POSITION FIXING

In position-fixing duties, reporting items change according to the following navigational conditions, so reporting must be done accordingly:

1) When navigating a straight planned course line
2) When navigating on a planned course including course alteration
3) When navigating toward a destination with a fixed arrival time

8.10 NECESSARY CONDITIONS FOR MOTIVATING TEAMWORK OBSERVED IN ACTUAL CASES

Various situations of team activities have been learned from BTM/BRM training. Cases observed during training and summaries of those conditions necessary for team motivation are introduced.

(1) **Inadequate understanding of necessary functions to achieve objectives**
This is a case observed during sailing on approach to a pilot station.
When navigating toward a pilot station, the ETA at the pilot station and the ship speed when picking up the pilot are important for maneuvering. To meet these requirements, the captain evaluates the remaining distance to the pilot station and the remaining time to arrival, and makes moment-by-moment adjustments in speed. Thought must be given to avoiding risks from other traffic vessels and the effects of wind and tide. Thus, the captain, in the role of team leader, considers all aspects, and adjusts speed for the purpose of satisfying arrival time and speed at arrival. It is important for the team members to achieve the functions required to support the busy captain. The moment-by-moment speed adjustments are required to arrive at the pilot station at the expected time. Thus, the remaining distance to the pilot station and the remaining time to arrival are necessary information, and since members assigned to position fixing are all aware of necessary information, the necessary speed can be calculated by the member assigned to position fixing. Typically, the average speed for adjusting arrival time is called *required speed*. Members assigned to position fixing must report course and distance to the pilot station at the time of position fixing. Thus, required speed also must be reported as an important function to further support the captain. However, in cases observed in training, there was no reporting of required speed despite reporting of course and distance to the pilot station. In training situations in which such cases arose, to maintain the expected time to the pilot station, speed was often rapidly reduced near the station, or the ETA could not be maintained.
What was missing that allowed this situation to arise? What must be improved so that the same situation may be avoided? First, members assigned to position fixing must be conscious of the scope of the tasks assigned to them, and perform all of them that are necessary. However, an important point indicated is that team leaders should observe the activity level of team members and always achieve the functions to promote members' activity needed for the team. The team leader had to request to be informed of required speed in this situation.
From this case, it is learned that all team members must have the same objectives and the same behavioral awareness, and team leaders

must make an effort to motivate teamwork. Communication and cooperation are basic functions of teamwork for all members, and the role of the leader should be mastered by the leader.

(2) **Insufficient achievement of assigned duties**

This case often arises due to the combination of constituent team members.

In BTM/BRM training, training was provided to three persons constituting a team: a captain, a second officer, and a third officer. A single helmsman was also assigned to the bridge, but this person was a staff member of the center and did not perform any duties required for navigation other than steering rudder. Training was conducted with this arrangement. In captain's briefings, assigned roles were announced by the captain. The second officer was to mainly execute position fixing and offboard communication, while the third officer was to mainly perform lookout, onboard communication, and engine telegraph operation. The second officer already had practical experience as a second mate, while the third officer had 1.5 years of onboard experience. With this team makeup, training was conducted in navigating the Singapore Straits. The second officer was also highly competent in communicating off ship in English, and skillfully achieved his assigned tasks in a short time. In intervals between assigned tasks such as position fixing and so on, he even managed visual lookout of surrounding ships and lookout through RADAR/ARPA, and made reports to the captain. The third officer was concentrating on detecting ships presenting risks mainly through RADAR/ARPA. Other ships can be detected quickly by visual lookout, which also quickly identifies ships presenting risks. Thus, the second officer was earlier in reporting on surrounding ships, which included the identification of ships presenting risks. For example, even when the third officer detected ships presenting risks early on with RADAR/ARPA, reporting on those ships would be quickly done by the second officer. Thus, the captain would ask the second officer for reports on surrounding ships. Consequently, the third officer was kept occupied only with RADAR/ARPA operation, which ended up being the same as doing no actual reporting. BTM/BRM activity was achieved by the captain and second officer throughout. In debriefing after the exercise, the third officer was asked why he did not report on the status of traffic vessels obtained through lookout. His response was "I did not know what to report".

The instructor thought that the inadequate reporting by the third officer was a result of his losing chances to report because the second officer had already reported before he tried. Certainly, the presence of particularly active team members frequently has the effect of reducing the actions of others. This indicates that the organization of constituent team members changes how they achieve their functions.

Furthermore, in the case introduced, the insufficient competency on the part of the third officer was a factor.

From this case, it is learned that all team members must achieve their assigned responsibilities whatever the circumstances. This is an essential point in constituent team members achieving maritime competency.

(3) **Insufficient communication between team members**

The case introduced here was also observed during BTM/BRM training, similarly to the previous section.

A team was formed of a captain and two navigators sailing westbound in the Singapore Straits. The second officer would obtain the information on traffic vessels in the vicinity of destinations from a marine traffic center via VHF wireless telephone and report it to the captain. The third officer was assigned the duty of looking out for traffic vessels using RADAR/ARPA. He initially detected a leaving ship from the Jong Fairway and reported it. Afterward, he was concerned about a risk presented by the leaving ship and continued to watch its movements. However, this leaving ship information had already been obtained by the second officer, and the target was gradually accelerating on a course ahead of their own ship westbound in the Singapore Straits from the Jong Fairway. The captain also had changed the course to port side, 5°, in response to this ship. The third officer did not hear the conversation between the captain and the second officer, so he determined that the target from the Jong Fairway was accelerating, would near their own ship and present a risk, and continued to report on the leaving ship's movements. Consequently, at the time when the target sailed ahead of their own ship, he said, "It looks like it is sailing ahead of our own ship, isn't it?" The captain and second officer then understood for the first time that the third navigator did not know the situation.

From this case, it is learned that all team members must share the same information. This is an essential point in cooperative activity.

(4) **Reported information not adequately reflected in maneuvering**

This case was seen during sailing of a very large crude carrier (VLCC) westbound in the Singapore Straits.

This situation was observed on a ship sailing through the waters around Buffalo Rock with its intended pass off Raffles Lighthouse. As they were passing through narrow waterways, the second officer aimed to perform position fixing every 5 minutes to verify ship position. Conditions were restricted with a visibility of 1 mile. The third officer was on lookout duties and reported that a ship sailing on an eastbound lane was closing to their own ship. The distance to the other ship was already closing to 1 mile. The captain immediately ordered the second officer by VHF communication to request that the other vessel take collision-avoiding action. However, the other ship was a VLCC headed toward the Shell single buoy mooring (SBM) and claimed priority of navigation. Thus, their own ship needed to engage in collision avoidance. The captain immediately gave a command for port 30° rudder angle, and then had the ship make a rapid turn to portside and continue sailing. Consequently, the ship entered the eastbound lane and then found itself in a risky situation with a new eastbound ship. While avoiding collision with the VLCC bound for the Shell SBM, the second officer had actually reported that they were closing the separation zone of the fairway, and the third officer had reported that they were closing the eastbound ship in the eastbound lane. The captain had concentrated on avoiding the immediate risk, and these reports had not reached him.

From this case, it is learned that the captain should make his or her final decisions by integrating the work of each team member. This is a necessary function of the captain.

(5) **Insufficient action by team members**

This case was observed during BTM/BRM training with a team consisting of three persons including the captain.

The third officer was on lookout duty. The ship being handled departed from Tokyo Bay with a restricted 1 mile visibility. The plan was to pass east of Izu Oshima Island, by a VLCC that would pass through the waters off Kannonsaki at 09:00, and travel westbound after changing course at the Ryuosaki Lighthouse on Izu Oshima Island. Around the time they would be leaving the Uraga Channel, they would encounter many large ships headed into Tokyo Bay. The next object for altering course would be Tsurugisaki Lighthouse. When the ship continued southbound at service speed, a report was received from the third officer on lookout duty that they would meet a succession of northbound ships. The captain continued maneuvering for collision avoidance while constantly verifying encountering situations. However, the CPA that could be achieved by avoiding the successive ships at close range would always be under 5 cables, meaning that a risky situation continued. At this time, the captain recognized that first detections of those ships presenting risks and the reports thereof were late. The captain requested the third officer to "perform early first detection and make reports promptly". The third officer noticed that the RADAR/ARPA range used was always 3 miles. As long as a close range was used, the detection of other ships would inevitably be late. Generally, when visibility is limited, the range of RADAR/ARPA used becomes smaller. Under orders from the captain, the monitoring range was thereafter widened, which performed early detection and enabled collision-avoiding action with sufficient time to spare.

From this case, it is learned that the captain must notice the behavior of constituent team members and constantly maintain optimum teamwork. This is a necessary function of the captain.

Through these cases, the conditions necessary for team motivation to achieve safe navigation are summarized.

Necessary Conditions to Motivate Teams

1) The team leader needs to make an effort to motivate teamwork.
2) All team members need to achieve their assigned duties.
3) All team members need to share the same information.
4) The captain integrates the results of work performed by each team member and then makes a final decision.
5) The captain needs to monitor the behavior of the constituent team members and constantly maintain optimum teamwork conditions.

What can be understood from the cases listed in detail is the importance of constituent team members fulfilling their functions. These include the *necessary functions of team leaders, adequate communication between team members, cooperative functions* to ensure smooth teamwork, and accomplishment of *assigned roles*. Effective teamwork can be achieved if team members familiarize themselves with these functions and then perform them with thorough understanding.

In this section, cases have been presented again to reinforce understanding of these functions. These cases should be kept in mind as a simple reference when engaging in team activities, and may also serve as an on-hand memo for effective performance.

8.11 EFFECTIVE USE OF RESOURCES

Finally, let us consider effective resource use. This is described here to deal with the fact that at present, the terms BTM and BRM are both used together by the IMO. Particulars on BTM and BRM have already been explained, and so the details are omitted here. This section focuses on discussions from the viewpoint of the use of resources.

Here, the resources used for safe navigation are considered as being divided into "material resources" and "human resources".

(1) **Use of material resources**

Material resources refers to those resources that contribute to safe navigation, excluding seafarers on active bridge duty. Typical examples of such resources include nautical instruments such as electronic chart display systems (ECDIS) and RADAR/ARPA; compasses and indicators of rudder angle, ship speed, and engine revolutions; documents and VHF wireless telephones. The general characteristic of these resources is that they provide information.

When handling a ship, information required for decision making is acquired from instruments and documents and then used after being converted to suit specific purposes. When this is done, the sources providing the information must be accurately understood. That is, where did the information originate from? Furthermore, what are the characteristics and properties of the information provided from those sources?

For example, there is a time delay when a status appears on an information display screen with ARPA. This characteristic or property of the information provided must be appropriately understood when using it. It would be a mistake to concentrate on only trying to improve competency in practicing techniques for ship navigation. Do not forget that improvement of competency is dependent on acquiring sufficient knowledge.

In general, information is provided by multiple instruments, and seafarers must make decisions based on the various kinds of information received. Currently, this feature tends not to be systematic, and in some cases there are overlaps with the same information provided by different equipment. There is no uniformity to the methods, and nothing has been systematized. This situation is largely related to the processes by which instruments are developed. Furthermore, there has been no adequate review of instrumentation overall, nor has there been any revision of regulations governing installed instruments. However, seafarers are currently required to maintain safe navigation under these present conditions. This is also a characteristic of the competency required of seafarers.

(2) **Use of human resources**

Next, we consider effective human resource use. The use of human resources largely falls on the captain in his or her role as team leader, but navigators also serve as leaders when on single watch. They must effectively use human resources such as the helmsman and able seamen (AB) (and function as a team).

A leader must also assess the abilities of team members and then constantly make estimates of their task progress and workload. This information must then be integrated into assigning and reassigning duties.

A leader also must communicate team objectives, that is, his or her intentions as leader, to members in different navigational situations, and provide advice and warnings so that the team is always active. Also, since he or she is also a team member, he or she must fulfill the following team member functions.

They must accomplish their own actions as a team member. The required behaviors for team members are summarized herein; they must complete the duties that have been assigned to them. These assigned duties must be completed fully. There is no such thing as an allowable lapse in duties. They are then required to employ communication to share information among other members on their intentions and the results of their work. In addition, they must cooperate in monitoring the progress of others' work and being monitored in return.

8.12 SUMMARY: NECESSARY COMPETENCIES FOR BRIDGE TEAM MANAGEMENT

The competencies that are the objectives of BTM are summarized as follows. The following competencies are the indispensable items required to maintain team activities.

(1) **Competencies required of team leaders**
   - The team leader clearly notifies all constituent team members of the details of objectives to be achieved by the team, and must specifically indicate the behavior concretely required to achieve those objectives.
   - The team leader must evaluate the competencies of constituent team members, and must provide clear commands regarding the duties assigned to all constituent team members and the details of those duties.
   - The team leader must constantly monitor the behavior of constituent team members, encourage team activities, and maintain optimum operational status. Since the team leader can judge the status of team member behavior based on the progress of assigned tasks and frequency of reporting, constant monitoring is required. When anyone shows inadequate behavior, the team leader must provide advice and commands to motivate them.
   - The team leader is also a member of the team, and thus must hold the following competencies required of team members.

(2) **Competencies required of team members**
   The competencies required of team members shown here are ones that must be fulfilled by team leaders as a member of the team.
   - Team members must communicate effectively so that all constituent team members share information.
   - Team members must constantly engage in cooperative behavior so that the behavior of individuals within the team enables smooth teamwork.
   - Team members must activate the overall actions of the team, and contribute to the achievement of team objectives, according to the intent of the team leader.
   - Team members must fulfill duties directed by the team leader.

Part II

Postscript

Management is a word that has come to be applied to many things in recent years. These include time management, financial management, system management, information management, and team management. Of the forms of management listed, the one that is most uncertain in its details but has received the most attention in recent years is *organizational management*. Considering that the English word corresponding to organization is "system", then system management and team management should be considered as clearly separate. In this book, *system management* is thought of as the various physical elements constituting a system; that is, the mechanical and engineering components. By contrast, *team management* is thought of as the functions required for a group of multiple individuals engaged in achieving a single goal to more efficiently and reliably achieve that goal, wherein team members are considered the constituent elements of an organization.

Thus, the subject handled in this book is the management of organizations consisting of two or more members. It was discussed what functions ought to be considered in order to achieve the goals set by multiple people in an organization. When the functions needed to achieve goals are simple, that is, if the tasks that can be done to achieve those goals are simple, then they may be accomplished by a single person. However, if the tasks are many and are also complicated, a single person cannot accomplish goals through his or her work. This corresponds to the situation faced by many companies and organizations in modern society that require the efforts of many personnel.

The author often explains the characteristics of teamwork through the following example. Here, let us imagine a grocery store. In small-sized stores, stocking, produce displays, pricing, customer service, and accounting are all done by a single shopkeeper. As the store's reputation grows and its merchandise gradually increases, as well as its customers, it develops into a supermarket that handles not only fruits and vegetables but various other products. A large store has many employees who must achieve goals for the business. Many departments are assigned to different functions, including stocking, advertising, sales, registers, and financial management.

Thus, an *organization* has been formed for achieving goals through the work of multiple employees. What might be the functions, that is, the work that must be divided among individual specialties and then accomplished so that the assigned departments serve their purposes? By thinking in this way, the functions that must be maintained for the organization to achieve its goals may be understood. The departments have responsibilities for stocking, advertising, and asset management. Now, the tasks performed by these individual departments are intrinsically important; however, they cannot be executed independently. What goods to stock is determined by demand; sales and advertising are based on stock in inventory; and sales methods must be devised. In other words, cooperation is required between all departments.

Interactive activities between assigned personnel are required when many related persons are working together toward achieving goals, whether it is a large or a small organization. The functions necessary when multiple persons are working jointly may be defined as the functions necessary for an organization to achieve its goals. These functions are what is described as team management.

In this book, the functions necessary for organized action to realize safe navigation have been explained. This part has been concluded by pointing out that the functions for achieving the goals of an organization mentioned here do not apply only to safe navigation, but can be broadly deployed in other social situations involving the work of organizations.

Part III

Bridge Team Management/ Bridge Resource Management Training

Chapter 9   Training System 189
Chapter 10 Bridge Team Management Training Structure 199
Chapter 11 Bridge Team Management Training Examples 211

Part III

Preface

In the following chapters, the details of a model course of bridge team management (BTM)/bridge resource management (BRM) training using a ship handling simulator are dealt with. The basic knowledge an instructor must possess before implementing training is explained. The purpose of training is to improve the ability to perform necessary techniques (defined as "competency"). A basic requirement of instructors is to accurately measure and assess trainee competency. The instructor must also understand how competency can be improved. Herein, the process for improving competency is explained.

Knowledge and competency should be considered as separate abilities for performing techniques. Knowledge is knowing what techniques are necessary, and competency is the ability to perform the necessary techniques. Safe navigation requires the competency to perform necessary techniques, but this is preceded by acquiring the relevant knowledge. Consequently, to realize safe navigation, knowledge must first be acquired, followed by the competency to execute it. Thus, if a seafarer has the knowledge but not the competency for appropriate use, their activity will not contribute to safe navigation. The author would emphasize the necessity of full competency.

Figure III.0.1 is a conceptual diagram that shows changes in BTM knowledge and competency level in a trainee during the course of BTM training.

The horizontal axis in Figure III.0.1 indicates the time elapsed during training, and the vertical axis represents the process of increasing knowledge and competency during the training course. Since this process starts from the origin point in the figure, this indicates that both knowledge and competency start from zero. However, this is not really the case, since there is generally a certain degree of knowledge and competency even at the initial stage of training.

The main thing that improves during the lectures held at the start of the training course is knowledge, whereas competency is usually not expected to greatly improve during lectures. Consider a case of planning to improve competency through exercises on a ship handling simulator after the relevant lectures are completed. In the exercises, trainees are assigned duties using the knowledge and competency they possess. Consequently, a valid

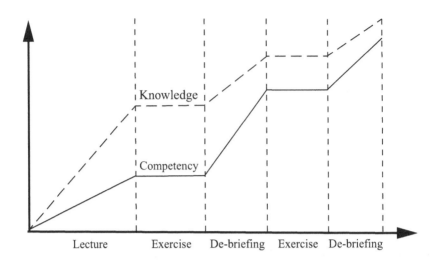

Figure III.0.1 Changes in knowledge and competency during progress of training

consideration is that knowledge and competency remain constant and do not improve during the exercises. That is, the purpose of exercises using a ship handling simulator is to determine the evaluation of trainee competency by creating an environment similar to navigating a real ship, having seafarers execute the actions they would do on a real bridge, and observing those actions.

The period during training when knowledge and competency improve is debriefing. During exercises using ship handling simulators, instructors observe the conversation and action of trainees and assess their competency on the basis of an evaluation list. This assessment is done to determine the degree to which trainees are fulfilling their roles as team members and achieving the functions necessary for the overall team as well as the functions necessary for team motivation. During debriefing, instructors provide the advice required for improving competency in order to achieve the necessary functions. Instructors judge whether trainee actions meet the recommendations or if there are any inadequacies. Whenever such inadequacies occur, instructors explain that such actions hinder safe navigation and clarify the correct actions and their necessity. When necessary, basic information is explained in the form of a lecture. During exercises using a ship handling simulator, debriefing is an important process for determining training effectiveness. The trainees must strive to accept, understand, and acknowledge that instructions during debriefings are necessary for competency improvement. Trainees must always verify unclear points with the instructors.

Chapter 9

Training System

In this part, specific methods of education and training for competency in bridge team management (BTM) are described. In Chapter 9, an educational system for developing competency and the purpose of developing BTM competency are explained. In Chapter 10, the structure of BTM training is explained in terms of the details of education and training and the program. The objective is to clearly understand the need for a logical training structure. In Chapter 11, examples of BTM/bridge resource management (BRM) training are introduced in order to understand a specific training implementation system. Through the explanation in this part, the system required for actual training implementation will be understood. In addition, implementation methods will be understood.

9.1 SUMMARY

The course introduced here is a model course for education and training developed on the basis of the competency necessary for BTM/BRM explained in Part II of this book. This model course is certified by the Ship Maneuvering Simulator Committee of the Japan Institute of Navigation (JIN) as "a standard model course for implementing education and training in BTM/BRM training by using a ship handling simulator".

This course comprises lectures, practice, and exercises using the ship handling simulator. Lectures are focused on explaining *the basic competency* necessary for safe navigation, *the significance of bridge teamwork, and necessary competency*. In addition, lectures are conducted on *passage planning* and on *ship maneuverability*. Passage planning is practiced using nautical charts of water areas that are navigating areas in the exercises on the ship handling simulator.

Prior to exercises with the simulator, familiarization training to understand the way to use the navigational instruments and methods of communicating with outside agencies is implemented on the simulator bridge under instructor guidance.

Exercises using the simulator are planned such that the navigation difficulty increases as the course progresses. In addition, *this training is implemented under navigational conditions that reveal the significance of bridge teamwork in different situations to achieve the necessary competency*. This includes navigation situations during pilot boarding, in addition to main engine and steering system malfunctions and abnormal behavior of other traffic ships.

Exercises consist of captain's briefings, exercises in navigation using the ship handling simulator, and subsequent debriefings by instructors on practical competencies.

Through this course, trainees acquire the necessary competency for BRM and BTM while on watch as well as learning the importance of bridge team activities and the actual specific competency for BTM.

9.2 OBJECTIVES OF EDUCATION AND TRAINING

The purpose of receiving the BTM training introduced here is to acquire the following competency required of seafarers. When a ship proceeds into situations in which navigational safety cannot be ensured by a single seafarer because of congested water areas, narrow waters, and/or rough weather, in other words very difficult navigational situations, seafarers require *specific competency to maintain safe navigation by multiple seafarers*. That is the objective of this training.

Trainees who complete this course learn the significance of bridge team activities and master the necessary competency as follows:

1) The necessary competency for multiple seafarers to cooperate on safe navigation
2) The importance of multiple seafarers performing assigned tasks
3) The necessity of cooperative individual actions when multiple seafarers are achieving tasks together
4) The importance of mutual understanding through essential communication in order to report the results of the assigned tasks and to share awareness and recognition of situations the ship being handled is facing when multiple seafarers are achieving tasks together
5) The importance of seafarers serving as the leaders of teams consisting of multiple seafarers understanding the importance of team leaders making necessary decisions and knowing what must be done to motivate teamwork

9.3 CONDITIONS NECESSARY FOR ACHIEVING THE OBJECTIVES OF BRIDGE TEAM MANAGEMENT TRAINING

In this section, the methods for effectively conducting BTM training are explained. Training is effective when trainees understand the purpose of training, master the necessary knowledge, and can take practical action based on this knowledge. Thus, trainers must always verify whether the intentions of trainees are heading toward the purpose of training and motivate them to make improvements on their own.

(1) Understanding the purpose of BTM

First, the trainee must understand that the purpose of BTM training is different from that of basic techniques training. The trainee must understand that the training for *basic techniques* focuses on the necessary competency for safe navigation while on single watch. Basic techniques are explained in detail in Part I of this book.

The main purpose of BTM training is to make trainees understand the necessity *to maintain team functions and to promote team activities*. This critical explanation may not be possible *if the difference between the competency needed when on single watch and the functions necessary during teamwork cannot be clearly understood*. One way of effectively addressing this is to use *the concept of elemental technique development*, since it explicitly lays out the necessary techniques for safe navigation. As described in Part I of this book, in elemental technique development, the functions necessary for safe ship navigation are broadly divided into nine elemental techniques. Of these, the first seven techniques are needed when on single watch: planning, lookout, position fixing, maneuvering, observance with traffic regulations, instrument operation, and communication. BTM training consists mainly of education and training in the ninth technique: management. Management techniques differ depending on what is being managed. The key point is that personnel in team member assigned to the bridge serve the function of promoting team activities: the primary objective of BTM.

During training, an effort must be made to ensure that the trainees can understand the purpose of BTM.

(2) Understanding that BTM is necessary for safe navigation

Trainees must simultaneously understand *both the purpose and the necessity of BTM*. A lack of understanding of this necessity may result in low motivation and lack of effort to achieve the purpose of BTM.

The necessity of BTM is easily understood when analyzing the results of accidents caused by lack of BTM. In many cases, cause analysis of accidents due to a lack of BTM reveals that even if necessary functions for ship navigation are being achieved by the entire team, the achievement level of necessary functions for safe navigation is lower than the normal level when seafarers are on single watch. In other words, for the overall team to achieve the necessary functions for safe navigation, *it must act as a team activity; in addition, the individual members must perform the basic navigation techniques required when on single watch.*

When the necessary techniques are performed by the team as a unit, the results of tasks assigned to its multiple team members are integrated to realize safe navigation.

The relevant techniques are those for basic navigation. These correspond to techniques ranging from executing assigned necessary functions, namely tasks, to achieving the results. However, a required function of teamwork is to reflect the assigned functions of team activities. It is incorrect to understand team activities solely as the execution of assigned tasks. Members joining in team activities based on such understanding cannot contribute the behaviors necessary for achieving the whole-team functions. The trainees must understand that performing only one's own tasks does not fulfill one's role as a team member. This understanding can act as a measure of whether the necessity of BTM has been understood by the trainee.

The necessary functions for BTM are described in detail in Part II of this book.

9.4 TRAINING SYSTEM

(1) **Entry standard**

The persons taking the course introduced here must be seafarers working on ocean-going vessels. They must all have minimum experience of 1 year onboard. Those taking the course as team leaders in particular must have minimum onboard experience of 6 years as well as experience as a chief officer or captain. This is because all participants must be able to achieve the competencies necessary for safe navigation.

The aforementioned experience may be on any type of ship and navigational route. However, preferably, the experience must be significant. This is because *the purpose of this course is for constituent team members to acquire competency in effective team activity.* Consequently, this training is not suited for participants who are not familiar with certain types of ship or navigation areas. Team members tend to be incapable of demonstrating their essential team activity abilities in circumstances that differ from their experience. In this case, it becomes difficult to evaluate trainee team activity competency. Thus, the purpose of this training is not to become perfect in knowledge of the types of ship and navigation areas in which trainees are inexperienced, but the understanding and acquisition of the functions necessary for team activity. Therefore, the entry standard must take into account the abilities and experience of participating trainees in order to be sure of achieving the purpose of training.

(2) **Issuing course completion certificates**

Persons who have taken this course and have been evaluated as having the prescribed competency are issued a certificate attesting that they have completed a BRM and BTM training course conforming to the criteria for the standard course specified for BTM/BRM education and training by the Ship Maneuvering Simulator Committee of JIN (June 2013).

Furthermore, if a trainee cannot achieve the prescribed competency level during the training period, the training period may be extended and their competency retested.

(3) **Number of participating trainees**

This course includes three sets of exercises using a ship handling simulator, with three to four seafarers receiving training in each round. Consequently, if six to eight trainees participate, then two teams are formed, requiring six sets of exercises on a ship handling simulator. These two teams of participants are called Group A and Group B. When Group A is performing exercises with the simulator, Group B observes their handling from a separate room. After each group has participated in simulator-based exercise and observation, all members of both groups participate in the debriefing and receive explanations from an instructor. When two groups receive training, both groups receive lectures and practice simultaneously, while exercises are done separately. Thus, the course period is extended.

This course provides training in effective team activity by organizing teams consisting of a minimum of three people. Consequently, if training involves five participants, then one of the participants receives training as part of both Group A and Group B for effective training.

(4) **Instructor qualifications**

Instructors must have competency to be capable of each of the following in relation to lectures, practice, and exercises using ship handling simulators:

1) Create appropriate training courses
2) Understand the necessary competency for BTM and BRM
3) Understand the necessary competency for achieving safe navigation
4) Conduct lectures scheduled as part of the course and appropriately answer questions
5) Understand criteria established for assessing competency, and evaluate trainees' competency on the basis of their behavior
6) Provide proper explanations according to trainee competency during debriefings after simulator-based exercises in a ship handling simulator

Instructors must complete instructor training courses held by the JIN Ship Maneuvering Simulator Committee in order to develop abilities for the aforementioned qualifications. Instructor training courses are held regularly; the schedules and other material can be obtained from the secretariat of the JIN Ship Maneuvering Simulator Committee.

The following lectures and practice are held during instructor training courses by the JIN Ship Maneuvering Simulator Committee:

Lectures

1) Systemization of techniques for ship handling and elemental technique development
2) Necessary technique of BTM/BRM
3) Methods of preparing a syllabus for education and training using a ship handling simulator
4) Preparation of teaching materials for education and training using a ship handling simulator
5) Methods of preparing scenarios for education and training using a ship handling simulator
6) Methods for assessing competency during education and training using a ship handling simulator
7) Methods of effectively conducting debriefings after exercises with a ship handling simulator
8) Methods of passage planning

Practice

1) Preparation of teaching materials for education and training using a ship handling simulator
2) Preparation of scenarios for education and training using a ship handling simulator
3) Preparation of competency assessment lists for education and training using a ship handling simulator
4) Passage planning
By completing this training, instructors can acquire the indicated abilities.

(5) **Personnel needed to implement training**

The following personnel with the prescribed abilities are needed to conduct this course:

- **Instructors:** One or more persons with instructor qualifications.
- **Simulator operators:** Two or more persons to operate the ship handling simulator during its use in simulator-based exercises and to support the instructor in order to achieve training objectives.

(6) **Education and training equipment**

This course requires the use of a ship handling simulator.

Furthermore, the ship handling simulator requires the following nautical instruments, which are usually installed on a ship, and equipment for controlling ship movements.

- **Information display devices:** Compass indicating heading direction, log indicating ship speed, rudder angle indicator, rate-of-turn indicator, engine revolutions indicator, propeller revolutions indicator, propeller pitch indicator for controllable pitch propeller (CPP)-equipped ships, relative wind direction/relative wind speed indicator, RADAR/ Automatic Radar Plotting Aids (ARPA), etc.
- **Communication equipment:** Telephone devices capable of onboard communication with all parts of the ship, very high frequency (VHF) radio communication devices for offboard communication, whistles, etc.
- **Ship movement control equipment:** Steering gear, engine control systems, side thruster control systems
- **Other:** Equipment for making nautical chart entries, such as nautical chart tables, navigation light control panels, and display devices

Ship handling simulators must accurately reproduce the movements of an actual ship. In addition, they must have visual field reproduction equipment that uses realistic images for viewing navigational conditions from the simulated bridge.

In addition, equipment for monitoring the behavior and conversation of trainees on the bridge during simulator exercises is recommended.

Ship handling simulators must also have functions for placing other traffic ships in water areas intended for training and for controlling their movements.

(7) **Educational materials**

In addition to this book, the following materials must be prepared and provided in order to conduct this course:

- Nautical charts and nautical chart tables
- Information charts on water areas being navigated during exercises in ship handling simulators
- Logbooks
- Audio and behavioral monitoring devices
- Audio and behavioral recording devices

PCs and other projection equipment for demonstrating the results of lectures and debriefings

Chapter 10

Bridge Team Management Training Structure

10.1 COURSE TIMETABLE

(1) **Education and training program**

This education and training course is based on the following program, but it may be extended depending on the abilities of the participants. However, it cannot be shortened.

First day

1) Summarization of competency necessary for safe navigation when on single watch, verification of practical competency of trainees, and lectures to achieve an understanding of deficient competencies.

2) Passage planning for maneuvers during bridge team management (BTM)/bridge resource management (BRM) training with a ship handling simulator to be performed from the second day onward.

Second day

1) Learning of necessary techniques for BTM/BRM in classroom. In addition, understanding of the importance of BTM/BRM by analyzing examples of accidents

2) Understanding of the maneuvering characteristics of the ship during BTM/BRM training with a ship handling simulator in classroom

3) Familiarization in instrument operation using a ship handling simulator in order to gain proficiency in the instruments on the simulated bridge

4) Use of a ship handling simulator to gain in-depth knowledge of the maneuvering characteristics of the ship during simulated maneuvers

5) After that, exercise on BTM/BRM training (1) with a ship handling simulator

Third day
1) Exercise on BTM/BRM training (2) with a ship handling simulator
2) Exercise on BTM/BRM training (3) with a ship handling simulator
3) Post-training questionnaire survey

Practical exercises with a ship handling simulator are planned such that navigational difficulty increases gradually over the training course. In addition, these exercises are implemented under navigational conditions that require bridge team activities and necessary competency. Cases are also included of navigation situations during pilot boarding, in addition to main engine and steering system malfunctions and abnormal behavior involving other traffic ships.

(2) Education and training methods and time required

One example of a program for implementing education and training in this course is shown herein. Even if there are slight differences in implementation, the standard times used for lectures and practice exercises with a ship handling simulator are the same as (or longer than) those in the schedule:

First day
1) Lecture: 2 hours
   Course description and necessary conditions for safe navigation
2) Lecture: 2 hours
   Necessary techniques for safe ship navigation
3) Lecture: 2 hours
   Methods of passage planning
4) Practice: 2 hours
   Prepare three passage plans on nautical charts for navigating during simulator-based exercises, training with a ship handling simulator from the second day onward. Planning is done by team members collecting necessary information and entering whatever is necessary to realize safe navigation on nautical charts according to passage planning by the team leader.

Second day
1) Lecture: 2 hours
   Background for organizing bridge teams
2) Lecture: 2 hours
   Necessary bridge team functions for realizing safe navigation
3) Lecture and practice: 2 hours
   Lecture on ship maneuverability and maneuvering methods, practice in manipulating instruments on the bridge, and maneuvering in a ship handling simulator
4) Simulator-based exercises: 2 hours
   Captain's briefing by team leader before exercise and debriefing by instructor after exercises with ship handling simulator

Third day
1) Simulator-based exercises: 4 hours
    Captain's briefing by team leader before exercise and debriefing by instructor after exercises with ship handling simulator
2) Simulator-based exercises: 4 hours
    Captain's briefing by team leader before exercise and debriefing by instructor after exercises with ship handling simulator
3) Post-training questionnaire survey: 30 minutes
    After training is completed, a questionnaire survey of trainees on their awareness of training techniques and their opinions and expectations of training methods in order to verify the effects of training and to improve training itself is conducted.

(3) **Method of evaluating competency**
    When this course is conducted, there must be concrete and quantitative evaluation of whether the intended degree of competency has been achieved. Such evaluations must always be made objectively, and not according to the subjective judgment of the assessors. Thus, when planning training, in each exercise scenario with a ship handling simulator, a competency assessment list matching progress through a scenario must be prepared, and the competency of participants must be evaluated accordingly.
    The intended competencies of the course are as follows:
1) **Necessary competencies of team leaders**
   - Team leader must clearly notify all constituent team members of the objectives to be achieved by the team and indicate specifically the behavior necessary to achieve those objectives.
   - Team leader must evaluate the competencies of constituent team members, and clearly give orders regarding assigned duties, and the details of those duties, to all team members.
   - Team leader must constantly monitor the behavior of constituent team members, motivate team activities, and always maintain optimum activity. The behavioral status of constituent team members may be assessed according to the degree of achieving assigned tasks and the frequency of reports, so these actions should be constantly monitored by the team leader. A function of the team leader is to give advice and commands in order to promote team functions whenever they are inadequate. The team leader must achieve this by ascertaining the status of team members and then judging whether the motivational actions are being taken as needed.
   - Team leader is also a member of the team and thus requires the competencies required of the team members.
2) **Necessary competencies of team members**
    The competencies required of team members shown here are the competencies that must be similarly satisfied by the team leader as well.

- Team members must communicate effectively in order for all constituent team members to share information.
- Team members must always demonstrate cooperative behavior in order for the behavior of individual persons within the team to smoothly facilitate teamwork.
- Team members must motivate the entire team in its work activities and contribute to achieving team objectives according to the intentions of the team leader.
- Team members must fulfill duties ordered by the team leader.

The aforementioned evaluation criteria are for assessing the competencies considered necessary and essential for maintaining team activities. These then require the creation of training scenarios such that actions for evaluation are repeatedly demanded during exercises in the ship handling simulator. Competency is thus assessed each time an evaluable action is demanded by the scenario. Appropriate evaluation is made possible through evaluation lists in which situations requiring actions subject to competency assessment are listed according to progress through a scenario. Thus, preparation of these lists is essential, and evaluation criteria must be discussed by assessors beforehand to avoid discrepancies.

For evaluation, the competencies required in situations are those that repeatedly appear in various situations included within a scenario. Consequently, the degree to which techniques are achieved can be quantitatively verified by tallying the results of evaluations after exercises in the ship handling simulator. An example of an evaluation list is shown at the end of this part. Referring to this, one may see that the same evaluation criteria appear repeatedly for different events.

The degree of achievement is calculated after each of the three sets of exercises with a ship handling simulator. By comparing the degree of achievement after each of the three sets of exercises, it is possible to quantitatively evaluate improvement in competency.

Debriefings are conducted on the basis of the results of the aforementioned competency assessments. Inadequate action during exercises with a ship handling simulator is then presented in order to improve trainee competency.

10.2 DETAILS OF EDUCATION AND TRAINING

1) Syllabus for education and training

Education and training consist of classroom learning, exercises with a ship handling simulator, and post-exercise debriefings. Classroom learning is accomplished through lectures and practice. Lectures emphasize explanations of the basic competency necessary for safe navigation, the significance of bridge team activities, and the competencies necessary to achieve it. The objectives of these lectures are an understanding of the difference between being on single watch and being on duty with a team of multiple seafarers, and an understanding of the necessary functions of team activities. In addition, lectures are also held on passage planning and ship maneuverability. Practice sessions are assigned to develop passage planning abilities using nautical charts of the water areas where trainees will navigate during exercises with a ship handling simulator.

During exercises with a ship handling simulator, the captain's briefing by the team leader, which is an important element in team activities and for achieving safe navigation, is done first. Actual exercises with a ship handling simulator then follow. The objective of this training is to evaluate trainees against the necessary competencies; thus, assessments of trainee competency are performed from the captain's briefing by the team leader, and it is evaluated whether the team leader explained to the team members all items necessary for safe navigation. Furthermore, it is evaluated whether team members fully understand the directions of the team leader and are able to actively engage in the confirmation and verbal communication necessary for safe navigation. Captain's briefings by team leaders are done immediately before starting exercises.

During practical exercises with a ship handling simulator, *instructors must monitor trainee behavior and conversations, give full attention to evaluating their competency,* and not interfere with spontaneous actions in any way. Thus, trainees will execute the usual actions required for ship navigation under environmental conditions similar to those on a real ship. Observing usual trainee behavior makes it possible to make accurate estimates and evaluations of their competency. Thus, the assessment of trainee competency through exercises must involve methods not based on evaluators' opinions and memories. Prior to the exercises, assessors must prepare assessment lists according to progress through a scenario. Assessment then can proceed according to details that have been specified for exercises with ship handling simulators. To further ensure that assessments are objective and not based on assessor opinions, assessment criteria must also be discussed between instructors at the scenario creation stage.

Debriefings following exercises with ship handling simulators are based on exercise assessments of trainee competency by instructors. Visual recordings of trainee behavior during exercise, navigation records, and the like are effective means of ensuring that trainees have properly received the instructions.

2) Details of lectures

During training for development of competency in BTM/BRM, lectures provide knowledge of the competencies and methods to achieve competency, which are the objectives of training. In this section, the content of lectures, the education and training methods, and the time required are explained.

First day
1) **Lecture:** Course description and necessary conditions for safe navigation
   a. **Course description:**
      – Explanation of program and training schedule
      – Meaning of this training course
      – Techniques for which competency is to be improved through this course
   b. **Necessary conditions for safe navigation**
      # Reference: Part I, Chapter 2, "Analysis of Techniques for Ship Handling" in this book
      – Factors related to achieving safe navigation
      – Environmental factors
      – Seafarer competency factors
      – Necessary conditions for safe navigation
      – Changes in environmental factors and changes in navigational safety
      – Changes in seafarer competency factors and changes in navigational safety
2) **Lecture:** Necessary techniques for safe navigation
      # Reference: Part I "Techniques for Ship Handling" in this book
      – Ship handling techniques listed in the International Convention on Standards of Training, Certification and Watchkeeping for Seafarers (STCW)
      – Organization of important techniques for ship handling
      – Development of elemental techniques for ship handling
3) **Lecture:** Methods of passage planning
      # Reference: Part I, Chapter 2, "Technique of Passage Planning" in this book
      – Preparations to make passage plans
      – Important points of passage planning
      – Passage planning for coastal and restricted waters
      – Methods of entering plans on nautical charts

4) **Practice:** Prepare passage plans on nautical charts used in exercises with ship handling simulators

Second day
1) **Lecture:** Background for organizing bridge teams
    # Reference: Part II, Chapter 8, "Bridge Team Management" in this book
    - Necessary conditions for safe navigation
    - Relationship of BTM and BRM
    - Necessity of BTM
    - Necessity of BTM training
    - Origin of team management
2) **Lecture:** Bridge team functions necessary for realizing safe navigation
    # Reference: Part II, Chapter 8, "Bridge Team Management" in this book
    - Reason for organizing teams
    - Necessary functions of team management
    - Significance of communication
    - Significance of cooperative activity: Cooperation
    - Effective use of resources
3) **Practice:** Cause analysis of accidents due to inadequate team management
4) **Lecture and exercise:** Lecture on ship maneuverability and maneuvering methods, and practice with operating instruments on bridge and maneuvering in ship handling simulator.
    a. **Lecture on ship maneuverability and maneuvering methods**
        - Representative data illustrating maneuverability
        - Methods of using maneuverability data for maneuvering
    b. **Methods of operating instruments on bridge**
        - Characteristics and methods of operating RADAR/ Automatic Radar Plotting Aids (ARPA)
        - Methods of operating very high frequency (VHF) and other communication equipment
        - Methods of operating steering gear and engine operation devices
        - Methods of operating other instruments on bridge
    c. **Maneuvering practice**
        The team leader conducts preliminary training for understanding the maneuvering characteristics of the ship the trainees will maneuver with the ship handling simulator. During preliminary training, an understanding of their own ship performance is gained through the actual experience of operating the main engine and rudder, so team leaders acquire competency that will be reflected in the subsequent exercises with the ship handling simulator.

In the interim, the team members practice with the instruments on the bridge in order to gain proficiency in operating them and familiarize themselves with offboard communication methods.

3) **Relationship of each lecture to the purpose of education and training**

In Chapter 9, "Objectives of Education and Training", the five items listed here were explained as the "significance of bridge team activities and the necessary competency" to be learned by completing this course:

1) The necessary competency for multiple seafarers to cooperate on safe navigation
2) The importance of multiple seafarers performing assigned tasks
3) The necessity of cooperative individual actions when multiple seafarers are achieving together on tasks
4) The importance of mutual understanding through required communication in order to report the results of individual assigned tasks and share awareness and recognition of situations facing their own ship when multiple seafarers are achieving tasks together
5) The necessity for seafarers serving as the leaders of teams consisting of multiple seafarers to understand the importance of team leaders making necessary decisions, and what must be done to motivate teamwork

The knowledge for each intended technique is explained in its respective lecture, but item 1) in the foregoing list is covered in the first-day lectures on "Necessary techniques for safe navigation" and "Methods of passage planning". Subsequently, item 2) is covered in the first-day lecture on "Necessary conditions for safe navigation", and items 3), 4), and 5) are covered in the second-day lectures on "Background for organizing bridge teams" and "Bridge team functions necessary for realizing safe navigation".

4) **Details of exercises**

BTM/BRM training using a ship handling simulator is implemented over 2 days, on the second and third days of training. The necessity of BTM/BRM and a knowledge base of the necessary techniques for achieving BTM/BRM are understood through lectures beforehand. Thus, the purpose of simulator-based exercises is for trainees to learn how to put their knowledge into action as competency.

The following points must be reflected when implementing exercises so that they are conducted effectively and for trainees to learn knowledge and competency.

Exercises with a ship handling simulator must include the following four activities:

1) Pre-briefings immediately before exercises with a ship handling simulator
2) Captain's briefings by the team leader
3) Exercises with a ship handling simulator
4) Debriefing immediately after exercises with a ship handling simulator

Three sets of exercises with a ship handling simulator are implemented over 2 days incorporating the four aforementioned activities as a set. The details that must be considered with each of the aforementioned activities are explained.

1) Pre-briefings immediately before exercises with a ship handling simulator

Explanations are given by instructors immediately before exercises indicating navigation departure points and expected arrival points on nautical charts. Additionally included are explanations of the weather and sea state when starting maneuvers, confirmation of on- and offboard communication methods, and if necessary, other actions that must be taken during navigation, such as communicating with marine traffic control centers and pilot stations, and making commands according to scenarios.

2) Captain's briefings by the team leader

Captain's briefings by the team leader during exercises with a ship handling simulator have an educational and training purpose. They are meant to evaluate how the leader has planned objectives for the team, whether members have been notified of such plans, and whether the functions that must be performed by members in order to execute plans have been clearly conveyed. At this time, the leader must verify passage planning on nautical charts, show intentions, verify relevant items, and provide the necessary explanations for team members. Leader competency is evaluated based on the process by which these tasks are performed. In addition, not only are leaders evaluated during this period, but so is the attitude of team members toward team activities.

a. The specific main assessing points for team leaders are
   - Has a reasonable plan for achieving objectives been prepared?
   - Have members been clearly apprised of the plans the leader has drafted?
   - Have roles been appropriately assigned to individual team members?
   - Have individual members been explained any precautions for when performing assigned tasks?
   - Have it been explained to individual members what is necessary for team activities?

b. **Assessing points for team members**
    - Are the leader's plans and the intention understood?
    - Have any necessary advice and warnings been given to the leader?
    - Are assigned roles understood, and have precautions when performing them been confirmed?

3) Exercises with a ship handling simulator

The purpose of exercises with a ship handling simulator is the assessment of trainee competency in BTM/BRM. Therefore, events requiring the achievement of such competency repeatedly appear in the training scenarios used. Assessors monitor trainees' behavior and conversations during each event, and then assess their individual knowledge and competency in BTM/BRM.

Exercises with a ship handling simulator require the reproduction of expected events in order to assess trainee competency, which is the responsibility of the instructors. Trainee actions and behavior can easily break situations of expected events. Therefore, instructors must plan on reproducing expected events and on constantly monitoring changes in those events while making assessments.

4) Debriefing immediately after exercises using a ship handling simulator

The positives and negatives of training effects are decided by the contents of debriefing. The effectiveness of training corresponds to the improvement of trainees' competencies and intention through training. The first issue is whether trainees recognize the importance of improvement in competency, which is the objective of training. To achieve the training objective, instructors must point to trainees' behavior if there were factually any inadequacies in the actions taken by a trainee that resulted in loss of safety. Furthermore, if there were inadequacies in trainee actions but it is factually unclear whether this resulted in a loss of safety during training, then an effective measure is to restate the fact from lectures that inadequate action increases the possibility of a loss of safety.

For the trainee to understand the instruction, an effective way to show behavior is by video recordings taken during exercises.

Getting all trainee competencies to the desired level during training is sometimes difficult because of time constraints. However, exercises using a ship handling simulator are implemented three times, so an effective means is to evaluate competency improvement as a trainee progresses through training, and then indicate the degree of improvement and the goal of improvement at debriefing.

Particularly during the initial exercises with the ship handling simulator, if many inadequacies are detected, it must be kept in mind that it is important for trainees to achieve competency finally. This means that not all inadequacies are indicated immediately, and the scope of improvements is increased step by step. Instructors need to remember that limiting the number of improvements instructed ensures that trainees remain focused on their goals.

Instructors must objectively assess trainee achievement in the competencies that are the objectives of training according to evaluation lists, and then systematically debrief trainees on the indications in the three sets of exercises.

The extent to which a trainee improves in technical competency is *strongly related to their desire to do so and their recognition of its necessity.* An effective means to improve trainee desire and recognition may be to present to them records of the operations they have performed, changes in ship movements, and the track of their ship.

The objectives of training are to educate the trainees and improve their competencies. Instructors must always be aware of these points and offer appropriate guidance.

Chapter 11

Bridge Team Management Training Examples

11.1 IMPLEMENTING TRAINING

Scenario development for exercises with a ship handling simulator is planned such that navigation difficulty increases gradually. In addition, exercises are implemented *under navigational conditions that highlight the significance of bridge team activities and the necessary competency*. This also includes maneuvering situations during pilot boarding, in addition to main engine and steering system malfunctions and abnormal behavior of other sailing ships, so the importance of the necessary competency for BTM/BRM is understood under various conditions.

The BTM/BRM education and training course presented here consists of three sets of exercises with a ship handling simulator. Herein, the exercise implementation plan, the methods of developing exercise scenarios, and the evaluation lists used are explained. The details can be found in the instructor training course held by the Ship Maneuvering Simulator Committee of the Japan Institute of Navigation (JIN).

(1) **Exercise implementation plans**

The BTM/BRM education and training course presented here consists of three sets of exercises with a ship handling simulator. In the initial practical exercise set, instructors observe the actions taken by trainees during the exercise, particularly the degree to which they demonstrate competency in team management, which is the purpose of this training. Instructors should determine what training is necessary to improve trainee competency levels related to team management. Furthermore, since the individual competencies of the multiple trainees participating in a team are not uniform, their individual competency levels must be identified and evaluated. Here, *training that suits trainee competency levels* means the selection of scenarios determined for the second set of exercises onward and methods for giving explanations during debriefings.

According to the aforementioned plan, debriefings are repeatedly held for the second and third sets of exercise with a ship handling simulator such that competency gradually improves to the intended level. However, since time is limited, it is not necessarily possible for all trainees to achieve improvement in all competencies during training. An effective way of dealing with such situations is to provide appropriate guidance at the final debriefing held after the third set of exercises and suggest technical goals for trainees to concentrate on while onboard a ship.

(2) **Scenario development guidelines**

The scenarios used in the three sets of exercises with a ship handling simulator must include multiple *situations for evaluating competency* in BTM/BRM, which is the purpose of this education and training course. The act of creating situations for assessing competency is called *the setting of events*. Required events must be developed in accordance with trainee competency levels. Thus, even though the actions taken by trainees are different according to the events, the competency being assessed is still BTM/BRM competency. Depending on the type of the event, the difficulty of performing BTM/BRM techniques changes. The difficulty of performing a technique corresponds to difficulty in achieving safe navigation, so events must be set during scenario development with a difficulty level suitable for the competency level of the person being trained. Thus, during the course of the three sets of exercises with a ship handling simulator, difficulty is increased gradually, and trainee competency improvement is evaluated.

The factors deciding difficulty are as follows:
1) Density of ships sailing in navigation area
2) Number of ships encountered during navigation
3) Conditions of ship encounters (number of ships presenting a risk)
4) Navigational area conditions such as geographic and water area conditions
5) Weather and sea state
6) Type of vessel being handled
7) Condition of navigational equipment available on bridge
8) Other factors affecting the content and quality of tasks performed by trainees

Difficulty is determined by the quantity and quality of the tasks performed. Therefore, by considering the quantity and quality of the aforementioned factors, and simultaneously occurring combinations of different factors, scenarios including various types of difficulty can be created.

(3) **Guidelines on competency assessment list development**

Competency assessment lists are developed to clarify the assessment of competency and evaluation of trainee behavior as a scenario progresses. The purpose of this training course is the development and improvement of trainee competency in BTM/BRM. Therefore, assessment items must correspond to items for evaluating the degree of achievement in BTM/BRM techniques. Even if the events contained in a scenario are different, the same assessment items are shared. However, the difficulty of achieving a technique differs, so the results of assessments differ depending on individual circumstances. Here, the evaluation items in assessments for three examples with a high level of difficulty in achieving the techniques are listed.

Assessment items corresponding to possible events

1) In *the event of main engine and steering system malfunctions*, actions following the detection of emergencies are important. All members are notified of the situation by the member who first detected it, including the leader. Subsequently, the leader considers corrective actions, and then notifies all members of the chosen actions. When this is done, an important requirement of members is whether they can give appropriate advice to the leader. Members must deal with the situation according to the commands of the leader, but they must also maintain their assigned functions for continuing safe navigation, such as lookout and position fixing. When a member taking action to deal with malfunctions is unable to fulfill his or her originally assigned functions, they must be handled by other members instead. Since normal duty arrangements are altered by the actions of each member, the leader must appropriately reorganize the role assignments while continuing to monitor the behavior of team members. The aforementioned actions are the subject of assessments of BTM/BRM competency achievement.

2) Assessment of competency *in situations with abnormal behavior on other ships* starts from detection of the abnormal behavior. As in the preceding paragraph, the member who first makes the detection notifies all members, including the leader, of the situation. While the leader considers appropriate action, the members verify nearby traffic vessels and safe water area conditions, and then inform all team members, including the leader. It is the role of team members to provide information and advice to support decision making by the leader. When specific members are charged with communicating with abnormally behaving ships and contacting nearby traffic vessels, the substitution of other members to perform functions is also an important assessment objective. Leaders must make appropriate decisions according to the situation. In addition, leaders must appropriately reorganize role assignments while maintaining the monitoring of team members' behavior.

3) Basic to maintaining team activities among bridge team members are the team activities *during maneuvering with pilots*. Normally, when a pilot is boarding, an extreme decrease in reporting by team members

is observed. During pilot boarding, the captain, acting as team leader, must effectively communicate with the pilot, verify the pilot's maneuvering plans, and verify the pilot's intentions during maneuvers. In addition, members must provide adequate information on lookout and positioning such that the pilot can safely maneuver. During pilot boarding, whether bridge team members have the competency to perform the aforementioned actions is subject to BTM/BRM competency assessment.

When there is a significant increase in the tasks necessary for safe navigation, there is a great need for communication, cooperative action, and team management competency on the part of the leader required for BTM/BRM. Such difficult situations may only confuse novices, which is something that must be avoided. However, it may be appropriate for trainees who have learned the basics of BTM/BRM to be subjected to highly difficult training.

11.2 EXAMPLES OF EXERCISES USING SHIP HANDLING SIMULATOR

The following materials used in practical exercises with a ship handling simulator are included at the end of this part.

1) Scenario summary (Table III.11.1)
2) Arrangement of ships presenting risks during navigation (Figure III.11.1)
3) Competency assessment list (Table III.11.2)
4) Example of tally results (Table III.11.3)

The scenario presented here is one developed for assessing actual BTM/BRM competency on a ship westbound in the Singapore Straits. BTM/BRM training must be repeated every 3 to 5 years and is required to maintain and improve the necessary competency. This scenario is one for trainees who have already received training two to three times, and was developed for first exercises with a ship handling simulator conducted on the second day of the training course. This training is conducted under the assumption that trainees have already learned the applicable competencies to a certain extent. In terms of content, this scenario is one of medium difficulty and is used to assess the degree to which trainees maintain the necessary competency based on their actions during training, and to discover what direction to take in terms of the education and training techniques employed in their subsequent later training. During debriefing after exercises with a ship handling simulator, trainees are informed of the present status of their competency maintenance and areas of inadequate competency.

The time required for this scenario is approximately 1 hour and 10 minutes. The supposed visibility is set according to trainee competency and includes both good visibility and restricted visibility conditions. The primary situation for achieving safe navigation is navigation while passing through a narrow waterway, while avoiding collision with ships crossing in various encounters.

The assessment list for this scenario is organized as follows.

First, the behavior of team members is assessed during the captain's briefing. The function section in the assessment list shown in Table III.11.1 contains the following:

L: Leadership (Assess the degree to which team leader functions are being achieved. Acting as team leader, all roles are assigned in a timely manner, including monitoring of team member behavior, motivating activity for them and reporting maneuvering intentions, etc.)
C: Communication (Communication, especially communication between team members, is an important assessment competency)
T: Teamwork (Team members are required to constantly act as members of the team. As mentioned under cooperative action, members must

achieve the necessary competency so that the team can constantly maintain necessary functions)

I: Instrument operation (Assess whether instruments are being effectively used in terms of BRM)

P: Procedure (Assess performance of basic techniques. This includes look-out, position fixing, observance of laws and regulations, etc.)

In the assessment list, the same evaluation items repeatedly appear for each event that arises. Thus, the characteristics of how trainees perform techniques are ascertained after exercise with a simulator by calculating the achievement of techniques in each of the aforementioned functions.

The page after the assessment list shows an example of a summary of assessment results. The results are for training in which three trainees have been assigned the roles of captain, second officer, and third officer. In the case of the captain, the degree of achievement is calculated for five functions, including leadership. In the case of the second and third officers, leadership is left out, and the degree of achievement in four functions is tallied. The degree of achievement for each item is indicated by the percentage of functions evaluated as achieved versus the number of evaluable functions. In addition, the overall percentage of functions achieved is calculated for each item.

In the calculation, communication, leadership, and teamwork are the competencies particularly targeted for improvement by this education and training course. Therefore, it is important to carefully examine the degree of achievement, explain it at debriefings, and use it in subsequent future training.

Table III.11.1 Scenario summary. Westbound in Singapore Strait

1. Initial Condition (Own ship)
   Initial Position: 01-13.7 N, 103-55.15 E
   Initial Course:<246>
   Eng. Telegraph: S/B Full Ahead, 12 knots

2. Weather Condition
   Wind: Calm
   Tidal Current: Nil
   Visibility: 1.5 NM

3. Scenario summary

Time	Event	Vessels concerned	Assessing behavior
00:00	Navigate on Singapore Strait westbound lane Check of surrounding situation	#9 Small cargo #3 PCC #18 Container	Confirmation on traffic vessels
00:05	#5 crossing from starboard side	#5 Container	Communication using VHF Avoiding collision
00:15	Fishing boats	#7, #8, #21, #26	Using horn for warning of risky situation Avoiding collision
00:20	Alter course at Batu Berhanti	Batu Berhanti	Control ship's motion
00:30	Merging vessel from Jong Fairway to Main Strait (westbound lane)	#19 Cargo vessel	Communication of risky situation using VHF Avoiding collision
00:40	Crossing VLCC heading to Shell SBM from eastbound lane	#17 VLCC	Communication of risky situation using VHF Avoiding collision
00:45	Alter course at Gusong light buoy	Gusong	Control ship's motion
00:55	Cargo vessel overtaking from port side	#2 Cargo vessel	Communication of risky situation using VHF Avoiding collision
01:00	Alter course at Raffles lighthouse and passing report to Singapore VTIS		Control ship's motion Communication to VTIS
01:05	Small cargo vessel crossing from starboard side	#24 Small cargo	Communication of risky situation using VHF Avoiding collision
	End		

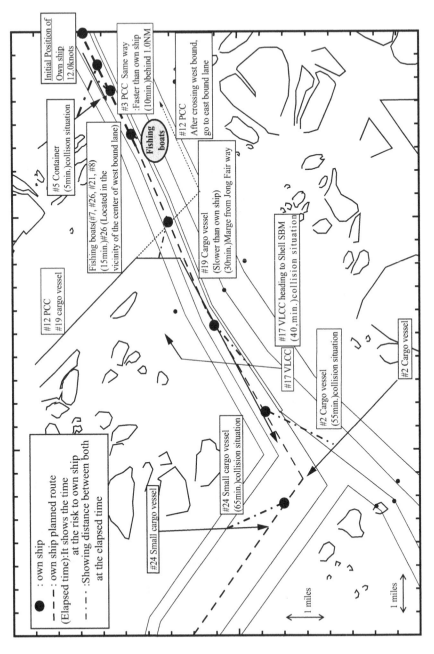

Figure III.1.1 Main events in scenario (westbound in Singapore Strait)

Table III.11.2 Competency assessment list. Assessment sheet on mariner's behavior

Event	Assessing behavior	Function	Capt. Name +1 Attain	Capt. Name 0 Insufficient	Capt. Name −1		2/O Name +1 Attain	2/O Name 0 Insufficient	2/O Name −1		3/O Name +1 Attain	3/O Name 0 Insufficient	3/O Name −1
Nav. Plan	Proper Nav. Plan is made	P	Attain	Insufficient	Lack								
	Communication on Nav. Plan is done	C	Attain	Insufficient		C	Attain	Insufficient		C	Attain	Insufficient	
Confirmation on vessel's surroundings													
#Cargo on head	Confirmation on situation is done	P	Attain	Insufficient		P	Attain	Insufficient		P	Attain	Insufficient	
#Small vessel on port bow	Checking Nav. instruments is done	I	Attain	Insufficient		I	Attain	Insufficient		I	Attain	Insufficient	
#PCC on stern	Communication on the situation and assigned tasks are done	C	Attain	Insufficient		C	Attain	Insufficient		C	Attain	Insufficient	
#Cargo	Assignments of all tasks are done	L	Attain	Insufficient									
	Attain the tasks and team activities	T	Attain	Insufficient	Lack	T	Attain	Insufficient	Lack	T	Attain	Insufficient	Lack
Crossing vessel													
#5 Container	Detection on target in early stage is done	P	Attain	Insufficient		P	Attain	Insufficient		P	Attain	Insufficient	
	Adequate usage of Nav. instruments is done	I	Attain	Insufficient		I	Attain	Insufficient		I	Attain	Insufficient	

(Continued)

Table III.11.2 (Continued) Competency assessment list. Assessment sheet on mariner's behavior

Event	Assessing behavior	Function	Capt. Name				2/O Name				3/O Name		
			+1	0	-1		+1	0	-1		+1	0	-1
	Communication with target is done (incl. VTIS)	C	Attain	Insufficient		P	Attain	Insufficient		P	Attain	Insufficient	
	Communication on the target and assigned tasks are done	C	Attain	Insufficient		C	Attain	Insufficient		C	Attain	Insufficient	
	Proper actions are done	P	Attain	Insufficient									
	Assignments of each tasks are done	L	Attain	Insufficient									
	Attain the tasks and team activities	T	Attain	Insufficient	Lack	T	Attain	Insufficient	Lack	T	Attain	Insufficient	Lack
Fishing boats #7,#8,#21,#26	Detection on target in early stage is done	P	Attain	Insufficient		P	Attain	Insufficient		P	Attain	Insufficient	
	Adequate usage of nav. instruments is done	I	Attain	Insufficient		I	Attain	Insufficient		I	Attain	Insufficient	
	Communication with target is done (incl. airhorn)	C	Attain	Insufficient		P	Attain	Insufficient		P	Attain	Insufficient	
	Communication on the target and assigned tasks are done	C	Attain	Insufficient		C	Attain	Insufficient		C	Attain	Insufficient	

(Continued)

Table III.11.2 (Continued) Competency assessment list. Assessment sheet on mariner's behavior

Event	Assessing behavior	Function	Capt. Name +1	Capt. Name 0	Capt. Name -1		2/O Name +1	2/O Name 0	2/O Name -1		3/O Name +1	3/O Name 0	3/O Name -1
	Proper actions are done	P	Attain	Insufficient		P	Attain	Insufficient		P	Attain	Insufficient	
	Assignments of each tasks are done	L	Attain	Insufficient									
	Attain the tasks and team activities	T	Attain	Insufficient	Lack	T	Attain	Insufficient	Lack	T	Attain	Insufficient	Lack
Altering course													
Sakijang or Batu Berhanti	Confirmation on the way point in early stage	P	Attain	Insufficient		P	Attain	Insufficient		P	Attain	Insufficient	
	Adequate usage of nav. instruments is done	I	Attain	Insufficient		I	Attain	Insufficient		I	Attain	Insufficient	
	Communication on the W.P. and assigned tasks are done	C	Attain	Insufficient		C	Attain	Insufficient		C	Attain	Insufficient	
	Proper actions on Alt Co. are done	P	Attain	Insufficient									
	Assignments of each tasks are done	L	Attain	Insufficient									
	Attain the tasks and team activities	T	Attain	Insufficient	Lack	T	Attain	Insufficient	Lack	T	Attain	Insufficient	Lack

Table III.11.3 Total score of achieving degree. Westbound of Singapore Strait (start from the northeast side of Batu Berhanti)

<CAPT>	A	Number of assessing item	Number of assessing item	Achieved degree
L: Leadership		9	9	100.0%
C: Communication		15	15	100.0%
T: Teamwork		9	9	100.0%
I: Instrument		9	6	66.7%
P: Procedure		17	16	94.1%
Total		59	55	93.2%
<2/O>	B	Number of assessing item	Number of assessing item	Achieved degree
L: Leadership		0		
C: Communication		10	9	90.0%
T: Teamwork		9	6	66.7%
I: Instrument		9	4	44.4%
P: Procedure		15	13	86.7%
Total		43	32	74.4%
<3/O>	C	Number of assessing item	Number of assessing item	Achieved degree
L: Leadership		0		
C: Communication		10	5	50.0%
T: Teamwork		9	5	55.6%
I: Instrument		8	3	37.5%
P: Procedure		18	9	50.0%
Total		45	22	48.9%

Postscript

Engineers can be broadly classified as those who create things and those who operate them so that they can fulfill the purpose for which they were made. The former utilize their knowledge to its fullest extent in order to create things that satisfy specific objectives. On actual sites, the latter operate those things that the former engineers have created in order to fulfill their purpose under certain environmental conditions. And, the greater the target of management becomes, the more the system needs to be composed of multiple elements, which include a wide variety of wisdom and knowledge.

The value of such large systems is not completed when it has been developed. Such value can be recognized only when the intended work and functions are achieved in practice. *Engineers on actual sites* make use of knowledge and wisdom to realize the intended work and functions.

Here, let us consider the engineers who handle the large structural systems that are ships: seafarers. When considering the conditions under which seafarers serve their functions in order to realize the primary goal of safe navigation, the same level of knowledge and wisdom for creating new systems is required. In large systems consisting of many subsystems, handling is not easy. Furthermore, in addition to the environments in which ships are placed, being directly affected by natural conditions, navigation through irregular marine traffic is required. Seafarers who handle ships in such environments are constantly forced to deal with new and uncertain external factors. Whether or not large systems that have been created can demonstrate their objective effects and functions is determined by how the systems operate practically. Thus, the importance of the techniques employed by seafarers must be emphasized.

Ship handling techniques have developed over a long period of time. And thus, the techniques involved have mainly been discussed empirically. Compared with those empiricisms, this book proposes a system based on analyzing, organizing, and integrating techniques for ship navigation that have been developed over time. Such systematizing has not been prominent in seafaring to date. The validity of a new theory of ship handling with seafarers as system operation engineers must be verified from various angles. And, the ideas introduced in this book have previously been verified without being challenged. Accordingly, the decision to write this book was made to provide a systematic approach to safe navigation.

It is hoped that this book will serve as a guide for novices learning ship navigation techniques and will be helpful to engineers onboard in better understanding the importance of their own techniques and functions and for teaching them to the next generation. Furthermore, this will be an opportunity for navigation systems designers to understand the functions of seafarers and their characteristics, and also for educators and researchers in ship navigation to delve into ideas presented in the book, creating an impetus for further study.

Finally, much of the reference literature used for this book consists of academic papers presented at relevant academic conferences, and the author apologizes to the readers for using literature that is difficult to obtain. These research papers provide a detailed discussion of the content discussed in this book. Further research is continuing, and when the occasion arises, it will be introduced.

The author concludes this book by acknowledging Prof. Atsushi Ishibashi and Prof. Akiko Uchino for their work and research at the Faculty of Marine Technology of Tokyo University of Marine Science and Technology and for their efforts in proofreading and preparing drawings for this book.

Dr. Hiroaki Kobayashi

Summary of Key Factors

KEY FACTORS OF INTRODUCTION

1. Standard seafarers exhibit almost the same behavior under the same conditions.
2. Safe ship navigation is influenced by environment conditions and the behavioral characteristics of seafarers.
3. To realize safe ship navigation, it is necessary to create an environment that takes into account the behavioral characteristics of standard seafarers.

KEY FACTORS OF SECTION 1.2: FACTORS AFFECTING NAVIGATIONAL DIFFICULTY

Navigational difficulty is influenced by the following factors:

1) Ship maneuverability
2) Navigational geographic conditions, such as form of navigable waters and water depth
3) Weather and sea state
4) Marine traffic
5) Rules of navigation
6) Onboard handling support system
7) Onshore navigation support system

KEY FACTORS OF SECTION 1.3: SHIP HANDLING COMPETENCY OF SEAFARERS

The ship handling competency of seafarers is influenced by the following factors:

1) Seafaring qualifications of seafarers
2) Actual onboard navigational experience
3) Degree of seafarer fatigue
4) Degree of seafarer stress

KEY FACTORS OF SECTION 1.4: CONDITIONS NECESSARY FOR SAFE NAVIGATION

1. Safe navigation necessitates a balance of "the competency required for the environment" and "the achievable competency of seafarers".
2. When "the competency required for the environment" is higher than "the achievable competency of seafarers", safe navigation is difficult to realize.
3. When the "competency required by environment" is lower than "the achievable competency of seafarers", the possibility of realizing safe navigation increases.

KEY FACTORS OF SECTION 1.5: TECHNIQUES NECESSARY FOR SAFE NAVIGATION

In this section, techniques and competency pertaining to ship handling are defined as follows.

- Techniques are functions necessary to achieve safe navigation.
- Competency is the ability to perform the necessary techniques.

KEY FACTORS OF SECTION 2: DEVELOPMENT OF TECHNICAL ELEMENTS FOR NAVIGATION TECHNIQUES

The necessary techniques for realizing safe navigation are categorized and organized into the following nine elemental techniques:

- Technique of passage planning
- Technique of lookout
- Technique of position fixing
- Technique of maneuvering
- Technique of observing rules of navigation and other laws and regulations
- Technique of communication
- Technique of instrument operation
- Technique of handling emergencies
- Technique of management: managing techniques and team activity

KEY FACTORS OF SECTION 2.1: TECHNIQUE OF PASSAGE PLANNING

The following functions must be achieved through the technique of passage planning:

1) Collect information for safe passage planning
2) Use collected information to realize safe navigation
3) Perform passage planning in accordance with the collected information
4) Determine when the navigation conditions require changing of plans and accordingly create new plans

KEY FACTORS OF SECTION 2.2: TECHNIQUE OF LOOKOUT

Functions that must be achieved using the technique of lookout are as follows:

1) Understanding of current conditions
   - Early detection of other ships
   - Types of target ships encountered
   - Movements of target ships encountered (position, course, and speed)

2) Understanding of conditions that the ship being handled will encounter in the future
   • Future conditions of target ships encountered (future position, future course, and future speed)
   • Risks of collision (CPA, TCPA, and BCR)

KEY FACTORS OF SECTION 2.3:
TECHNIQUE OF POSITION FIXING

Functions that must be achieved by the technique of position fixing are as follows:

1) Select the method of information collection for position fixing.
2) Estimate the ship position on the basis of the collected information.
3) Evaluate the ship movement and estimate the information necessary to evaluate the effect of external disturbances and to realize passage planning.

KEY FACTORS OF SECTION 2.4:
TECHNIQUE OF MANEUVERING

Functions that must be achieved by the technique of maneuvering are as follows:

1) Measure and ascertain current movements of the ship.
2) Select instruments to operate in order to realize the planned ship motion.
3) Determine control variables of control systems to realize the planned motion.

KEY FACTORS OF SECTION 2.5: TECHNIQUE OF OBSERVING RULES OF NAVIGATION AND OTHER LAWS AND REGULATIONS

Functions that must be achieved using the technique for rules of navigation and other laws and regulations are as follows:

1) Understanding the relevant laws and regulations
2) Reflection and implementation of the laws and regulations in actual navigation

KEY FACTORS OF SECTION 2.6: TECHNIQUE OF COMMUNICATION

Functions that must be achieved by technique of communication are as follows:

1) Select methods for communication
2) Understand how to engage in communication
3) Understand when to engage in communication
4) Understand how to apply proper language

KEY FACTORS OF SECTION 2.7: TECHNIQUE OF INSTRUMENT OPERATION

Functions that must be achieved by the technique of instrument operation are as follows:

1) Understand the available instruments.
2) Understand the methods of using instruments to obtain necessary information.
   The following constitutes necessary information:
   • Information on traffic vessels
   • Information on the ship positioning for safe navigation
3) Understand the characteristics of information provided by the instruments.
4) Understand the methods of using the provided information.

KEY FACTORS OF SECTION 2.8: TECHNIQUE OF HANDLING EMERGENCIES

Functions that must be achieved by the technique of handling emergencies are as follows:

1) Identify location of problems.
2) Repair problems and malfunctions.
3) Identify and complete necessary tasks related to abnormal occurrences.
4) Detect and react to abnormal behavior in other traffic vessels.
5) Identify and respond to abnormal weather and sea state.

KEY FACTORS OF SECTION 2.9: TECHNIQUE OF MANAGEMENT

The targets of management in this book are human resource organization and techniques. The techniques necessary for managing human resource organization are described in Part II (Bridge Team Management). The key factors for technical management are summarized in the following text.

Functions that must be achieved by the management technique are as follows:

1) Select techniques that must be applied.
2) Select concrete functions of techniques to be performed.
3) Determine frequency and timing of performing techniques.
4) Determine priority when multiple techniques must be applied.

KEY FACTORS OF SECTION 3.1: POLICIES FOR IMPROVING THE COMPETENCY OF INEXPERIENCED SEAFARERS; FUNCTION ACHIEVED BY TECHNIQUE OF PASSAGE PLANNING

1) To estimate and confirm the situations faced during watch duty before standing watch
2) To enter necessary behaviors for the estimated situations marked on the charts

KEY FACTORS OF SECTION 3.2: POLICIES FOR IMPROVING THE COMPETENCY OF INEXPERIENCED SEAFARERS; FUNCTIONS ACHIEVED BY THE TECHNIQUE OF LOOKOUT

Functions that must be achieved by the lookout technique are as follows:

1) Primarily through visual observation, scan the lookout area both near to and far from the ship being handled.
2) Detect other traffic vessels as early as possible.
3) Collect the following information on the status of traffic vessels:
   • Distance and bearing of other ships to the ship being handled
   • Course and speed of other ships
   • Confirmation of risks to the ship being handled in terms of CPA, TCPA, and BCR.
4) Prioritize warnings to other ships on the basis of risks to the ship being handled.
5) Continuously monitor ships presenting risks, initiate VHF communication, and implement collision-avoiding actions in a timely manner where appropriate.

KEY FACTORS OF SECTION 3.3: POLICIES FOR IMPROVING THE COMPETENCY OF INEXPERIENCED SEAFARERS; FUNCTIONS ACHIEVED BY THE TECHNIQUE OF POSITION FIXING

Functions that must be achieved by the technique of position fixing are as follows:

1) Improvement is needed not only in position fixing using instruments such as ECDIS and GPS but also in competency in position fixing using cross-bearing and radar.
2) The purpose of position fixing is not solely to determine the present position of the ship being handled. The following information must be collected for safe navigation:
   • Measuring direction and distance to next waypoint
   • When a destination has a predetermined ETA, in addition to the direction and distance to the destination, the required speed to realize the ETA
   • Estimations of wind and tide affecting ship movements and their estimated effects on ship movement

KEY FACTORS OF SECTION 3.4: POLICIES FOR IMPROVING THE COMPETENCY OF INEXPERIENCED SEAFARERS; FUNCTIONS ACHIEVED BY THE TECHNIQUE OF MANEUVERING

Functions that must be achieved by the technique for maneuvering are as follows:

1) Understand the differences in the use of rudder commands and course commands as commands to helmsmen when altering course.
2) Fully understand ship maneuvering performance and issue appropriate steering commands when altering course.
3) Avoid small repeated heading angle changes when making course alterations for collision avoidance.
4) Be careful when making large course changes in order to avoid collisions with a third ship or deviating from the planned course line.

KEY FACTORS OF SECTION 3.5: POLICIES FOR IMPROVING THE COMPETENCY OF INEXPERIENCED SEAFARERS; FUNCTIONS ACHIEVED BY TECHNIQUES FOR OBSERVING LAWS AND REGULATIONS

Functions that must be achieved by techniques for observing laws and regulations are as follows:

1) Understand traffic laws related to navigation and behave in accordance with the laws and regulations.
2) Understand the scope of applicable laws and regulations.
3) Understand laws and regulations applicable to the planned navigation area beforehand, and discuss them with senior seafarers.

KEY FACTORS OF SECTION 3.6: POLICIES FOR IMPROVING THE COMPETENCY OF INEXPERIENCED SEAFARERS; FUNCTIONS ACHIEVED BY THE TECHNIQUE OF COMMUNICATION

Functions that must be achieved by the technique of communication are as follows:

Communication on the bridge is detailed in Part II of this book. The following are the essential points for communicating with other traffic vessels.

1) Clearly decide the purpose of communication and organize what is being communicated before establishing contact.

2) Ensure that sufficient time will be available for completing collision-avoidance maneuvers after the communication. In particular, communication for collision avoidance should be done as early as possible.

3) Master the basic English phrases essential for maritime communication as early as possible. It is particularly important to organize the issues to communicate in order to engage in a coherent conversation.

KEY FACTORS OF SECTION 3.7: POLICIES FOR IMPROVING THE COMPETENCY OF INEXPERIENCED SEAFARERS; FUNCTIONS ACHIEVED BY THE TECHNIQUE OF INSTRUMENT OPERATION

Functions that must be achieved by the technique of instrument operation are as follows.

The purpose of using navigation instruments is to obtain effective information for navigation. Seafarers must have a strong intention to use the information for safe navigation.

1) When using RADAR/ARPA:
   - Timely switching of radar range and observation of both long- and short-range targets
   - Setting motion vector lengths of other ships corresponding to the radar range
   - Using parallel indexing and BCR information and applying it to maneuvering
   - Attaining competency in prioritizing ships presenting risks and collecting the necessary information
2) Organizing information obtained from instruments installed on the bridge, understanding how to utilize it, and applying it to navigation.

KEY FACTORS OF SECTION 3.8: POLICIES FOR IMPROVING THE COMPETENCY OF INEXPERIENCED SEAFARERS; FUNCTIONS ACHIEVED BY THE TECHNIQUE OF MANAGEMENT

Functions that must be achieved by the technique of management are as follows.

An example of insufficient competency seen in inexperienced seafarers is competency in technical management. Competency in technical management requires the following:

1) Proper selection and timely performance of techniques: it is necessary to know when and how to perform techniques in navigational situations requiring the execution of multiple techniques.
2) When multiple elemental techniques must be performed simultaneously, the necessity and priority of the functions must be established.

The technique of management on techniques cannot be achieved without fully understanding the function of each elemental technique and how to perform it. Therefore, competency in each of the elemental techniques must first be improved.

KEY FACTORS OF SECTION 4.1: SIGNIFICANCE OF ELEMENTAL TECHNIQUE DEVELOPMENT

The following are made possible by developing elemental techniques for ship handling:

1) By analyzing safety in ship navigation based upon necessary techniques, it becomes clear that the attainment of necessary techniques determined by environmental conditions is a necessary condition for realizing safe navigation.
2) The techniques necessary for safe navigation in any situation can be organized into nine technical elements, and safe navigation can be achieved by synthesizing these nine elemental techniques.

KEY FACTORS OF SECTION 4.2: TECHNIQUES AND COMPETENCY

The following are made possible by clarifying seafarer competency:

1) By comparing the required competency level in a function against the actual performable function level of seafarers, seafarer competency can be clearly evaluated.

2) By clarifying the relevant techniques of implementing efficient and effective competency training, a logical training system can be created.

3) The required competency level differs depending on the navigation situation. Seafaring qualification ranks are determined according to the required competency level. Thus, the difficulty of a navigational condition is evaluated according to the competency level required to ensure safe navigation in that environment.

KEY FACTORS OF SECTION 4.3: COMPETENCY LIMITS AND COMPETENCY EXPANSION

The following points arise if it is assumed that seafarer competency is limited:

1) Seafarers have limitations in the level of their competency to achieve necessary techniques.

2) The limits of seafarer competency must be known in order to find the real causes of accidents. In addition, the conditions for attaining competency must be known.

3) Navigation instruments become effective when they enhance seafarer competency to achieve functions.

4) The development of the navigational environment must be based on the ability of the seafarer to achieve the necessary functions.

5) The safety of a navigational environment can be quantified by using standard seafarer competency as a measure.

KEY FACTORS OF CHAPTER 5: NECESSARY TECHNIQUES FOR SAFE NAVIGATION

The following techniques must be possessed in order to achieve safe navigation:

1) The nine elemental techniques must be executed completely for achieving safe navigation.

2) Each of the elemental techniques has its functions to be achieved.

3) Competency to achieve the necessary techniques may change the achievement level due to the effects of environmental factors.

KEY FACTORS OF CHAPTER 6: NECESSARY CONDITIONS FOR SAFE NAVIGATION AND BRIDGE TEAM

The following conditions must be understood to achieve safe navigation:

1) The achievable competency of seafarers and the competency required by the environment are related in order to achieve safe navigation.
2) The achievable competency of seafarers must be the same as or higher than the competency required by the environment in order to ensure safe ship navigation.
3) A bridge team must be organized when safety cannot be ensured by a single seafarer.
4) The purpose of organizing a bridge team is that the competency realized by a team of seafarers is higher than that of a single seafarer.

KEY FACTORS OF CHAPTER 7: BRIDGE TEAM MANAGEMENT

The following items can be learned from the background and the presented examples of bridge team management (BTM).

1) Management includes two ideas: BTM and bridge resource management (BRM).
2) BRM consists of the functions that must be fulfilled by the team leader as part of BTM. Consequently, BRM is considered a part of BTM.
3) Analyses of past accidents have shown that inadequate action by team members puts the entire team at risk of an accident.
4) All team members must achieve functions in order to realize all team activities.

KEY FACTORS OF SECTION 8.3: NECESSARY FUNCTIONS OF TEAMWORK

In order to achieve the purpose of organizing a bridge team with plural seafarers, team members must achieve the following functions:

1) The team leader must achieve the function of motivating the team so that it can achieve its purpose.

2) Team members must communicate effectively.
3) Team members must maintain cooperation in order to maintain smooth team activity.
4) Team members must properly perform the work assigned to them.
5) The team leader must achieve items 2 to 4 as a member of the team. Team members must also follow the intentions of the team leader.

KEY FACTORS OF SECTION 8.4: SIGNIFICANCE OF COMMUNICATION

The following functions are possible through proper communication:

1) Sharing of results by each team member, summarizing of obtained information, and determination of a teamwork plan.
2) Sharing of a sense of purpose by team members.
3) Detection of human error by a team member can break the human error chain.

KEY FACTORS OF SECTION 8.5: SIGNIFICANCE OF COOPERATION

The following functions can be achieved through cooperation:

1) Maintenance of relationship in work performed by each team member and smooth teamwork.
2) Substituting or supplementary work for other team members.
3) Monitoring of actions by other team members and detection of human error.

KEY FACTORS OF SECTION 8.6: FUNCTIONS OF TEAM LEADER

1) The team leader must *clearly notify all team members of the purpose* to be achieved by the team and present the *specific actions necessary* to achieve the purpose.

2) The team leader must evaluate the competency of team members and *provide clear instructions of the duties and specifics of work* to all constituent members.

3) The team leader constantly *monitors the actions of team members* and *motivates* the team to *maintain constant optimum activity.*

4) The team leader is also a member of the team, and thus must provide good *communication* in order that team members can understand the situation the team faces, and achieve *cooperation* in order to smoothly proceed with teamwork.

KEY FACTORS OF SECTION 8.9: METHODS OF COMMUNICATION

The following points should be considered when communicating with team members:

1) Ensure clear awareness of the purpose of communication
2) Classify information and items to be communicated
3) Select the timing for communication
4) Decide on the order of information to be communicated
5) Decide on the frequency of communication

KEY FACTORS OF SECTION 8.9: METHODS OF COMMUNICATION FOR LOOKOUT DUTIES

In lookout duties for collision avoidance, reporting items change at each of the following steps, so reporting must be done accordingly:

1) When detecting ships with risk of collision
2) During continuous monitoring of ships with risk of collision
3) Right before starting collision-avoiding action
4) During collision-avoiding action
5) At the end of collision-avoiding action

KEY FACTORS OF SECTION 8.9: METHOD OF COMMUNICATION FOR POSITION FIXING

In position-fixing duties, reporting items change according to the following navigational conditions, so reporting must be done accordingly:

1) When navigating a straight planned course line
2) When navigating on a planned course including course alteration
3) When navigating toward a destination with a fixed arrival time

Index

A

Accident cause investigation, xviii–xix
Achieving safe navigation, 121–130
Automatic Identification System (AIS), 9, 38, 43
Awareness
  as determining factors of seafarer competency, 125–126
  levels, 13

B

Basic techniques training, 192
BCR
  as affecting factors on lookout, 38
  for communication for lookout duties, 170–171
  for instrument operation, 63
  for performing lookout, 43–44
Behavioral characteristics, xv, xix–xxiv
Bow crossing range, see BCR
Bridge Resource Management (BRM), 114, 132–133
Bridge Team Management (BTM), 114, 132–134, 147–181

C

Captain's briefing, 166–167, 207–208
Closest point of approach, see CPA
Cockpit Resource Management/ Crew Resource Management (CRM), 137
COLREG (International Regulations for Preventing Collisions at Sea), 55, 87

Communication, 28, 30, 57–60
  in competency assessment list, 216
  methods for reporting, 168–173
  as necessary functions for BTM, 153, 154–158, 165, 178
  timing and details with other vessels, 64–65, 89–90
Competencies required of team leaders, 181, 201
Competencies required of team members, 181, 201–202
Competency, 5, 17, 20, 101–104
  level, 10, 19
  method of evaluating, 201–202
  necessary level, 102–103
  required for the environment, 15
  seafarer (see Seafarer competency)
Competency assessment, 102–104
  list, 201, 212–213, 215–216, 219–221
Competency of bridge team
  mean, 129
  necessary, 191
Conditions essential for seafarers, 25–26
Conditions necessary for safe navigation, 15–16, 97
Contingencies, 34
Cooperation, 159–161, 165, 178
Cooperative behavior, 152, 159–161
Countermeasures for preventing accident xix–xxii
CPA, 43, 62

D

DCPA, 44, 62
Debriefing, 188, 207–209

Developing competency, 189
    of BTM, 165
    of BTM/BRM, 204
Developing navigational instruments
    needs oriented, 106
    seeds oriented, 106
Difficulty, 6
    average, 7, 9–10
    navigational, 7–11
Distance at closest point of approach,
    see DCPA

E

Electronic Chart Display Information
    Systems (ECDIS)
    for lookout, 40
    for passage planning, 37
    for position fixing, 48, 63–64, 83
Elemental techniques, 25, 30
    development, 25, 30, 97–100
Emergency, 29–30, 68–70, 72
Estimated time to arrival (ETA), 48,
    52, 83
Exercises with ship handling simulator,
    187, 208, 215
External forces, 48, 54

F

First detection, 40–43, 45
Function, 5
    necessary for team leader, 162–164
Functional approach, 19

G

Global Positioning System (GPS), 47

H

Human resources, see Resources

I

IMO, 17, 19
Implementation plan, 211
Influencing factors, 27–29; see also
    Elemental techniques
Information centers, 33
Instructor, 195, 203–204
Instrument operation, 29–30, 61–67,
    91–93
    in competency assessment list, 216

International Convention on
    Standards of Training,
    Certification and
    Watchkeeping for Seafarers,
    see STCW
International Maritime Organization,
    see IMO
International Regulations for
    Preventing Collisions at Sea,
    see COLREG

K

Knowledge and competency, 187–188

L

Leadership, 215
Length of motion vector, 62; see also
    Relative motion; True motion
Lookout, 27, 30, 38–46, 81–82
    limits and developing navigational
    instruments, 106–107

M

Management, 25, 29–30, 71–77
    team, 29, 72–75, 118
    technical, 29, 71–72, 74,
        94–95, 118
Maneuvering, 28, 30, 50–54, 85–86
Marine traffic centers, 59, 80, 89
Material resources, see Resources
Measurement range, 62

N

Newly developed instruments, 64
Nine elemental techniques, 24–30;
        see also Elemental techniques
No.10 Yuyo Maru, xvi–xxiii
No-go area, 33, 37

O

Observing laws and regulations, 28,
    30, 55–56, 87–88
Offset center, 63

P

Pacific ares, xvi–xxiii
Parallel indexing

for emergency treatment, 70
for passage planning, 37, 80
for position fixing, 48, 64, 92
Passage planning, 27, 30–37, 63, 80
for emergency treatment, 70
Position fixing, 27, 30, 47–49, 83–84
in STCW, 18
Procedure, 216
Proportional integral differential,
    see PID

R

RADAR, 47, 83
range, 91
RADAR/ARPA(Automatic
    Radar Plotting Aids), 64,
    66, 91–92
Relative motion, 26, 44, 62
Required speed, 83
Resources, 179–180
human, 132, 179–180
material, 132, 179

S

Safe navigation, 15–16, 97–98,
    127–128
Safe ship navigation, xxiv
Safe waters, 33
Scenario development, 212–213
Seafarer competency, 12–14, 15,
    24, 101
limits of, 105–107
Seamanship, xv, 109
Significance of bridge team
    activities, 191
Simulator operator, 196
Standard seafarers, xix, xxii–xxiv,
    19, 24

STCW (International Convention
    on Standards of Training,
    Certification and
    Watchkeeping for Seafarers),
    19, 20, 23, 26, 115, 132
Support systems, xxi, 9, 24, 39
Surrounding ship density, 42

T

TCPA, 41, 43, 45, 82
Team leader, 132–133, 152–153, 207
Team management, *see* Management
Team member, 132–133
Team motivation, 174–178
Teamwork, 152–153
in competency assessment list, 215
Technical management, *see*
    Management
Techniques, 5, 17, 20, 101–104
navigation, 26–30
necessary, 17–18, 98–100
required for navigation
    environment, 12
for ship handling, 23
Time to CPA, *see* TCPA
Traffic control centers, 69
True motion, 44, 62

U

Under-keel clearance (UKC), 34, 37

V

Vector length, 91
Very high frequency, *see* VHF
Vessel traffic information service
    (VTIS), 27, 38
VHF, 57–58, 64, 72

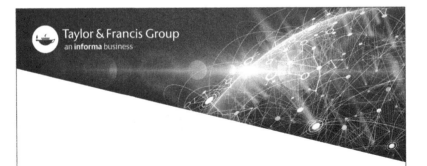

Printed in the United States
by Baker & Taylor Publisher Services